CHANGING RURAL LANDSCAPES

Edited by

Ervin H. Zube

and

Margaret J. Zube

CHANGING

The

University of

Massachusetts

Press

Amherst 1977

RURAL LANDSCAPES

Dedicated to Brinck, who speculates on, questions, and ponders the changing rural landscape.

CONTENTS

INTRODUCTION

Ever since the first census in 1790, American urban areas and their surrounding suburbs have grown, almost without exception, at a faster rate than rural areas. However, as first evidenced in the 1970 United States census, a dramatic shift has occurred in this pattern of growth and settlement—a shift unprecedented in modern America except for a brief period during the depression of the 1930s: the growth rate of rural counties now exceeds that of metropolitan counties. This rural population growth has been occurring not among the farm population, however, but in the small rural cities and towns. The farm population continues to decline.

The reasons for the shift are not yet completely clear. However, industrial relocations and dissatisfactions with urban life have been recognized as factors contributing to the growth of rural towns and cities. Lower taxes and large numbers of unemployed women in some rural areas have proved an attraction for labor-intensive manufacturing firms. The opportunity to escape from the noise, crime, and other perceived environmental and social disamenities of urban areas has attracted retirees, families, and segments of the youth culture. A major factor contributing to the decline in farm populations is the technological revolution in agricultural practices. With its potential for higher yields and its more limited use of manpower, mechanization has had considerable impact on the size and locus of the farm population.

As a consequence, these forces have brought about major changes in land use and in the look of the land. Some observers of these changes have expressed concern that much of the farm land which has been lost to production has been prime agricultural land—land which has since reverted to pasture and forest or has been converted to urban uses, highways, or other nonagricultural purposes. It is estimated that the net loss of this crop land reaches 1¼ million acres annually. These observers voice equal concern over the fact that this loss occurs at the same time that marginal lands are brought into production, including, for example, some 400,000 acres each year which require supplemental water for irrigation.

For several decades prior to this shift in settlement pattern, highly mobile Americans were also bringing about changes in the rural landscape in order to satisfy their ever-increasing demands for outdoor recreation. Summer and winter vacation complexes were developed, man-made lakes for water-oriented activities were created, and highways were built to move considerable numbers of people between urban and rural America.

The net impact of these and other forces of change, and their implications for the future of the *human* landscape have, with a few notable exceptions, been slow to be recognized. For a long time public policy continued to focus on agricultural production and land management rather than on broader questions and issues related to the rural landscape. Prior to the passage of the Rural Development Act of 1974, makers of public policy had given little attention to the stresses attendant on the growth and changing character of rural towns and cities, large-scale recreation developments, rural industrial development, and the social and environmental problems associated therewith.

Because of the complexity of the forces responsible for these changes, their recognition and study within a cultural, social, and historical context is important. One of the few persons who has had the vision to see this need is John Brinkerhof Jackson who, as author, lecturer, and editor of *Landscape* magazine (1951–1968), has been a guiding force in the study of the human landscape for a quarter of a century.

J. B. Jackson, as he became known to his readers, is a native New Englander. Born in Connecticut and educated at Harvard, he adopted the American Southwest as his home shortly after World War II. Well known among geographers, planners, architects, and landscape architects, he established a reputation as a keen observer of the American landscape and as a provocative and personable lecturer. In addition to frequent lecture appearances at universities across the country, he spends several months annually teaching and lecturing at both the Carpenter Center for the Visual Arts at Harvard University and the College of Environmental Design at the University of California at Berkeley.

Jackson shares with the geographers and landscape architects at Berkeley a long-standing interest in the premodern folk landscape of America. However, his primary focus in lectures and in journal and book contributions has been on the post-Civil War American landscape understood from the vantage point of native roots and values rather than from historical European standards of taste. From this perspective, Jackson's lectures and writings offer a positive outlook on the American landscape and what Americans have made of it.

In the spring of 1951 Jackson started *Landscape*—a journal which proved to be a thoughtful, provocative, and challenging forum for the discussion of the changing rural landscape. While the original focus was on the human landscape of the American Southwest, within six years it broadened to include human landscapes around the globe.

The philosophy of the journal reflects the philosophy of the man—exploratory and speculative in outlook. In the very first issue Jackson suggests to his readers that asking certain kinds of questions is more important than finding answers. He invites his readers to partake of a great adventure: studying the habitats and activities of man, viewing the world which we have helped to make, and viewing the changes which occur. "Wherever we go, whatever the nature of our work, we adorn the face of the earth with a living design which changes and is eventually replaced by that of a future generation. How can one tire of looking at this variety or of marveling at the forces within man and nature that brought it about?" As the first article ("The Need of Being Versed in Country Things") in the first issue suggests, the emphasis is rural, ". . . beyond the last streetlight, out where the familiar asphalt ends." Yet the urban landscape receives fair and measured consideration as well. Urban and rural are not separable; their influences—one upon the other—receive careful study.

Jackson's writings and those of his contributors are rarely out-of-date. They tend to question, probe, and illuminate processes and foster under-

standing of man's modification of the human landscape. Change is a recurrent theme both in *Landscape*'s "Notes and Comments" and in its
articles by Jackson and distinguished international contributors. Viewing
their subjects from both historical and contemporary perspectives, these
authors not only touch upon the larger cultural, social, economic, and
technological processes which modify the way in which the landscape is
organized and used, but they also present perspectives on such physical
manifestations of change as domestic architecture, artifacts (such as
fences and tombstones), the highway strip, the mechanization of farming,
and spatial organization of the land.

Along with the study of landscape change, contributors to *Landscape*
share a concern for the quality of the environment, not in terms of quan-
titative scientific assessments of air and water, but from a humanistic
point of view. The conviction persists that some landscapes are better
than others, not just more beautiful or efficient, but better from the
point of view of persons living in them. This approach, which considers
how man perceives and experiences the landscape, which considers man
as a part of the landscape, and which asks what it offers the individual,
is related to the basic philosophy of the journal. As stated by the editor
in the tenth-anniversary issue: "We are beginning to learn that the world
surrounding us affects every aspect of our being, that far from being spectators of the world, we are participants in it. Nature, therefore, is not
merely greenery; it is also the way we respond, physically and emotion-
ally, to greenery."

This volume is organized around a number of recurrent themes found in
the pages of *Landscape.* It consists primarily of Jackson's work, aug-
mented with complementary selections by other contributors. The dominant focus of the volume is change as it occurs in the rural landscape.
Each article, in its own way, deals with some aspect of this change.

The first section addresses some of the dominant cultural and technological forces that have brought about or influenced change both in the
premodern and the modern landscape. The authors suggest approaches to
the analysis and interpretation of the landscape with particular attention

to process, to human values, and to historical perspective. In the opening piece, Jackson sets forth a point of view which occurs throughout the book—that is, if we are to use the richness of our land wisely, we must plan carefully and well.

Another recurrent theme in *Landscape* is that of the spatial organization and relationships among elements of the landscape or, as Zelinsky calls them, "molecules of the human geographer's universe." The second section, "Territories and Boundaries," includes articles on some of the smaller of these elements such as front yards, walls, and fences, and discusses the effect of these elements on spatial organization and use. These elements are visual and cultural clues which enable us to read the human landscape and to discover the relationships between town and country, between past and present, or even—as suggested by Jackson's "Spring: Silent or Raucous?"—among man, birds, and land. The aggregation and interaction of these rural landscape elements or molecules combine to make up the rural community.

The third section of the book considers alternative community forms, the role of rural towns and cities, and the quality of rural community life. The interdependence of rural and urban is implicit in several of the articles, notably "Cities of the Plain" by Jackson and "What's the Use of Small Towns" by Solnit. The articles by Riley and Solnit also pose interesting and sometimes contrasting perspectives on the future of small rural communities.

The fourth and final section consists of three articles which reflect Jackson's continuing concern with a person-based approach to understanding the landscape. These articles offer provocative insights into perception and experience as ways of understanding the values individuals associate with landscapes. Such insights are important to the tasks of carefully planning the ever-changing rural landscape.

PROCESSES AND VALUES

What was to city dwellers, until a generation ago, merely a holiday venture into the rural world has become an earnest and almost desperate search for privacy and health and peace. — J. B. Jackson

The cultural landscape of the United States today expresses more clearly than ever the ambiguity of values inherent in our heritage.
— Philip Wagner

The kind of modification which the modern farmer undertakes is entirely different; different not only in scale but in purpose; for its purpose is to create an entirely new and artificial setting for his work. The ultimate aim is a man-made topography, a man-made soil, a man-made crop, all part of a new production process. — J. B. Jackson

There is plenty of room in our countryside for both kinds of landscape—the farmer's landscape of work, the city dweller's landscape of play, and there is plenty of talent among environmental designers to produce them. So far we have been singularly timid in America in the creating of artificial landscapes for recreation: the so-called English park is about as far as we have dared to go. But there is one lesson we can learn from the modern farmer: how comparatively easy it is to engineer the environment. There are few technological limits to our capacity to transform the land to suit our needs; all that is lacking is imagination and a sense of purpose. — J. B. Jackson

BACK TO THE LAND

J. B. Jackson

A high official in the Department of Agriculture recently (1957) had some remarks to make on a familiar topic: the changing status of the American farmer. After enumerating the many inventions which had transformed farmwork and telling us how country people had whole-heartedly adopted urban standards, he ended by saying that the line dividing the farm from the suburbs had all but vanished. Farming has ceased to be a way of life and has become a way of making a living.

Well, we have heard this before, and even the last sentence merely suggests the kind of change which has overtaken us, and the consequences of the change.

These, however, turn out to be far from discouraging. It is true that many of us will be sorry to admit that an ancient and highly respected calling has practically disappeared from our society, and that the average American farmer aspires to what is essentially a semi-white collar job in the suburbs. But that is mainly because we cling to certain traditional concepts. We have believed that the countryman belonged in a class by himself; we have believed that the rural landscape was exclusively his in a way the city was never the exclusive property of the townsman. It has seemed to us that his work must have provided unique joys and insights, and that it brought with it a special, almost mystical knowledge of the living world around him. And so we have generally assumed that farming had spiritual rewards which more than compensated for its hardships. Yet it is now quite clear that however the farmer of other regions may feel, or

[from *Landscape*, vol. 6, no. 3, Spring 1957]

the farmer of America may have felt in the past, the contemporary American farmer is more than willing to forego these intangible benefits for the sake of a better living and a steadier income.

No one will begrudge him the choice; but his flight from the land does not necessarily prove that we have been mistaken in our convictions. For even while the farmer is moving into town to commute to work in his fields, urban Americans are leaving the city in constantly growing numbers. And what are they looking for if not precisely those values which the commercial-minded farmer has chosen to abandon? What was to city dwellers, until a generation ago, merely a holiday venture into the rural world has become an earnest and almost desperate search for privacy and health and peace.

By an unparalleled stroke of good fortune the countryside has begun to fall vacant just as this flight from the city assumes impressive proportions. Just at the moment when our urban civilization seems to have reached the end of its spiritual resources it miraculously acquires an inexhaustible source of strength, and acquires it by default. Thus for the first time in history all classes of men can now experience the country environment—not, as in the past, in terms of ignorance and drudgery and exile, but in terms of joy, and of recreation in its truest sense.

But the acquisition brings with it many responsibilities which cannot all of them be delegated. The urban American must learn how to use this wealth, how to plan its development and preservation, and before he can establish a genuine claim to it, he must learn to love the land unselfishly and intelligently.

HOMESTEAD AND COMMUNITY
ON THE MIDDLE BORDER

Carl O. Sauer

The date of the Homestead Act, 1862, marks conveniently for our recall
a moment of significance in the mainstream of American history, the
great Westward movement of families seeking land to cultivate and own.
This started from states of the Eastern seaboard, swelled to surges across
the wide basin of the Mississippi-Missouri and ebbed away in the High
Plains. The Middle Border, as it has been named appropriately, was the
wide, advancing wave of settlement that spread over the plains south of
the Great Lakes and north of the Ohio River, making use of both water-
ways as approaches. Its advances made Cleveland, Toledo and Chicago
northern gateways. At the south, it gave rise to border cities on rivers,
such as Cincinnati on the Ohio, St. Louis at the crossing of the Mississippi, and Kansas City on the great bend of the Missouri. The Mississippi was
crossed in force in the 1830s, the Missouri River into Kansas in the border troubles prior to the Civil War. Although it did not begin as such, this
became the peopling of the prairies, the founding and forming of the actual Midwest.

The Homestead Act came pretty late in the settlement of the interior.
Land had been given free of cost to many. It had been sold at nominal
prices and on easy terms by public land offices and by canal and railroad
companies. The squatter who settled without title was generously protected by preemption rights and practices that grew stronger. Many millions of acres had been deeded as homesteads before the Act and many

[from *Landscape*, vol. 12, no. 1, Autumn 1962]

more continued to be acquired by other means afterwards. Land was long available in great abundance. The price in money of the wild land was the least cost of making it into a farm. Public land offices were set up to get land into private hands quickly, simply and cheaply.

The Indian Legacy

The American settler acquired learning that was important for his survival and well-being from the Indian, mainly as to agricultural ways. The settler was still a European in culture who had the good sense to make use of what was serviceable to him in the knowledge of the Indians of the Eastern woodlands. This learning began at Jamestown and Plymouth and was pretty well completed before the Appalachians were crossed.

Little seems to have passed from the Indians of the interior to the settlers. The Indian culture west of the Appalachians was still significantly based on cultivation, more largely so than is thought popularly to have been the case. Whether the western Indians contributed any strains of cultivated plants had little attention until we get much farther west, to the Mandans of the Upper Missouri and the Pueblo tribes of the Southwest.

Dispossessed of title to home, deprived of their economy and losing hope that there might be another start, many Indians were reduced to beggary or lived as pariahs about the white settlements. Their debauch was completed by alcohol, a thing wholly foreign to their ways, which became for them a last escape. Objects of despair to one another and of contempt and annoyance to the whites, the time was missed when the two races might have learned from each other and have lived together.

Most of the earlier American pioneers of the Mississippi Valley came by a southerly approach. They were known as Virginians and Carolinians, later as Kentuckians and Tennesseans, and in final attenuation as Missourians. They came on foot and horseback across the Cumberlands and Alleghenies, usually to settle for a while in Kentucky or Tennessee and thence to move on by land or river and cross the Ohio and Mississippi rivers. The relocations of the Lincoln and Boone familes are familiar examples.

7 Homestead and Community

The First Wave

Theodore Roosevelt hailed the main contingent of pioneer-settlers as Scotch-Irish, Mencken stressed their Celtic tone and temperament, Ellen Semple saw them as Anglo-Saxons of the Appalachians. Whatever their origins, and they were multiple, these were the backwoodsmen who brought and developed the American frontier way of life. They were woodland farmers, hunters and raisers of livestock in combination, and very skilled in the use of axe and rifle. Trees were raw material for their log cabins and worm fences, and also an encumbrance of the ground, to be deadened, burned or felled. The planting ground was enclosed by a rail fence, the livestock ranged free in woods or prairie. When the New Englander Albert Richardson reported life in eastern Kansas in the time of Border Troubles (*Beyond the Mississippi*) he said he could tell the home of a settler from Missouri by three things: the (log) house had the chimney built on the outside and at the end of the house; the house was located by a spring which served for keeping food in place of a cellar, and one was given buttermilk to drink. He might have added that there would be corn whiskey on hand and that if the family was really Southern the corn bread would be white.

This colonization was early and massive, beginning by 1800 and having the new West almost to itself until into the 1830s. At the time of the Louisiana Purchase, American settlers already held Spanish titles to a million acres in Missouri alone, mainly along the Mississippi and lower Missouri rivers. Their homes and fields were confined to wooded valleys, their stock pastured on the upland prairies. Nebraska alone of the mid-continent remained almost wholly beyond the limits of their settlement.

Viewed ecologically, their occupation of the land was pretty indifferent to permanence. Trees were gotten rid of by any means, the grasslands were overgrazed, game was hunted out. They were farmers after the Indian fashion of woods-deadening, clearing, and planting, and made little and late use of plow or wagon. The impression is that they gave more heed to animal husbandry than to the care of their fields or to the improvement of crops. Central and Northwest Missouri for example, the best flowering of this "Southern" frontier, developed the Missouri mule early in the Santa Fe trade, and later bred saddle and trotting horses and beef cattle.

The Wave From The North

The great northern immigration set in in the 1830s and depended from the beginning on improved transportation, the Erie Canal, steamships on the Great Lakes, stout and capacious wagons. It continued to demand "internal improvements," the term of the time for public aid to communications, first canals and soon railroads, only rarely constructed and surfaced roads. Wagon transport, however, was important and a wagon-making industry sprang up in the hardwoods south of the Great Lakes. It may be recalled that the automobile industry later took form in the same centers and by using the same skills and organization of distribution. Canals, most significantly the Illinois and Michigan Canal completed in 1848, linked the Great Lakes to rivers of the Mississippi system for shipping farm products to the East. Railroads were first projected as feeder lines to navigable waters. The first important construction, that of the Illinois Central, was chartered in 1850 to build a railroad from Cairo at the junction of the Ohio and Mississippi rivers to La Salle on the Illinois and Michigan Canal and on the Illinois River. It was given a grant by Congress of two million acres of land.

This last great movement of land settlement was out on to the prairies and it differed largely in manner of life and kind of people from the settlement of the woodlands. It depended on industry and capital for the provision of transportation. It was based from the start on plow-farming, cast iron or steel plows to cut and turn the sod, plows that needed stout draft animals, either oxen or heavy horses. By 1850, agricultural machinery had been developed for cultivating corn and harvesting small grains, and was responsible for the gradual replacement of oxen by horses as motive power.

The prairie homestead differed from that of the woodlands, in the first instance, by depending on plow, draft animals and wagon. It, too, grew corn as the most important crop, in part for work stock but largely to be converted into pork and lard by new, large breeds developed in the West that were penned and fed. Fences were needed, not to fence stock out of the fields but to confine it. The livestock was provided with feed and housing. The farm was subdivided into fields, alternately planted to corn, wheat, oats, clover and grass, arranged in a rotation that grew the feed for the work animals and for the stock to be marketed. A barn was necessary for storage and stabling. This mixed economy, its cash income from animals and wheat, spread the work time through the seasons and maintained the fertility of the land. It was a self-sustaining ecologic system, capable of continuing and improving indefinitely, and it was established by the process of prairie settlement. There was no stage of extractive or exhaustive cultivation.

By the time of the Civil War—in a span of twenty years or so—the prairie country east of the Mississippi, the eastern half of Iowa and northern Missouri were well settled. Some counties had reached their highest population by then. My native Missouri county had twice its current population in 1860. More people were needed to improve the land and to build the houses and barns than it took to keep the farms going. Some of the surplus sought new lands farther West, much of it went into building up the cities. These people who settled the prairies were farmers, born and reared out of the Northeast or from overseas, first—and in largest numbers—Germans, and thereafter Scandinavians. They knew how to plow and work the soil to keep it in good tilth, how to care for livestock, how to arrange and fill their working time. They needed money for their houses and barns, which were not log but frame structure with board siding. The lumber was mainly white pine shipped in from the Great Lakes, long the main inbound freight. They needed money as well as their own labor to dig wells and drain fields. The price of the land, again, was the lesser part of the cost of acquiring a farm. The hard pull was to get enough capital to improve and equip the homestead and this was done by hard labor and iron thrift. This is a sufficient explanation of the work ethic and thrift habits of the Midwest, often stressed in disparagement of its farm life. In order to have and hold the good land, it was necessary to keep to a discipline of work and to defer the satisfactions of

ease and comforts. The price seemed reasonable to the first generation
who had wrested a living from scant acres in New England or to those
who had come from Europe where land of one's own was out of reach.

The End of the Village

Dispersed living, the isolated family home, became most characteristic of
the "Northern" folk on the frontier. In Europe nearly everyone had lived
in a village or town; in this country the rural village disappeared or never
existed. Our farmers lived in the "country" and went to "town" on business or pleasure. The word "village," like "brook" was one that poets
might use; it was strange to our western language. Land was available to
the individual over here in tracts of a size beyond any holdings he might
ever have had overseas. The village pattern was retained almost only
where religious bonds or social planning prescribed living in close con-
gregation.

Normally the land holding was the place where the family lived and
this identification became recognized in the establishment of title. The
act of living on the land occupied was part of the process of gaining possession. As time went on, prior occupation and improvement of a tract
gave more and more weight to preemption rights; living on the land protected against eviction and gave a first right to purchase or contract for
warranty of ownership. The Homestead Act was a late extension of the
much earlier codes of preemption, by which possession by residence on
the land and improvement could be used to secure full and unrestricted
title.

The General Land Survey established the rectangular pattern of land
description and subdivision for the public domain. Rural land holdings
took the form of a square or sums of squares, in fractions or multiples of
the mile square section of land. The quarter section gradually came into
greatest favor as the desired size of a farm and became the standard unit
for the family farm in the Homestead Act. Thus four families per square
mile, a score or so of persons, were thought to give a desirable density of
rural population. The reservation of one school section out of the thirty-
six in a township, for the support of primary public schools, provided an

incentive for the only kind of public building contemplated in the disposal of public lands. Four homes to the square mile, and about four schools to the six-mile square townships, gave the simple general pattern for the rural geography of the Midwest. The pattern was most faithfully put into effect on the smooth upland prairies. Here the roads followed section lines and, therefore, ran either north-south or east-west, and the farmsteads were strung at nearly equal intervals upon one or the other strand of the grid. It is curious that this monotony was so generally accepted, even a clustering of homes at the four corners where the sections met (and giving the same density) being exceptional.

Little attention was given to the site where the house was placed or to the assemblage of the structures that belonged to the farm. The choice of location was of a good deal of importance, as, for instance, in exposure to wind and sun.

The logistics of home location is an attractive and hardly investigated field of study, as is, indeed, the whole question of the rural landscape and its changes. The location of house and farm, cultural preferences of different colonizing groups, microclimatic drainage and sanitation was unrecognized, the toll paid in typhoid and "summer complaint."

Building was starkly utilitarian and unadorned. Neither the log cabin of the woodlands nor the box-shaped frame house of the prairies, nor yet the sod house of the Transmissouri country (made possible by the sod-cutting plow) was more than compact and economical shelter, varying but little in each form. Ready-cut houses, of standard simple patterns, were already offered by railroads to buyers of their land, an early form of tract housing. Quality of house and quality of land seemed to be in no relation. The embellishment of the home and the planting of the yard were left mostly to the second generation, for country town as well as farm. The history of the dissemination of ornamental trees and shrubs might be revealing, perhaps to be documented through the nurseries that sprang up from Ohio to Nebraska.

The economy, from its beginnings, was based on marketing products, but it also maintained a high measure of self-sufficiency. Smokehouse, cellar and pantry stored the food that was produced and processed on the farm. The farm acquired its own potato patch, orchard, berry and vegetable garden, diversified as to kind from early to late maturity, for different flavors and uses, selected for qualities other than shipping or precocious bearing. The farm orchards now are largely gone and the gardens are going. Many varieties of fruits that were familiar and appreciated have been lost. A family orchard was stocked with diverse sorts of apple trees for early and midsummer apple sauce, for making apple butter and cider in the fall, and for laying down in cool bins in the cellar to be used, one kind after another, until the russets closed out the season late in winter. The agricultural bulletins and yearbooks of the past century invited attention to new kinds of fruits and vegetables that might be added to the home orchard and garden, with diversification not standardization in view. Exhibits in the county and state fairs similarly stressed excellence in the variety of things grown, as well as giving a prize for the fattest hog and the largest pumpkin.

The Self-Sufficient Family

The Mason jar became a major facility by which fruit and vegetables were "put up" for home use in time of abundance against winter or a possible season of failure in a later year. The well-found home kept itself insured against want of food at all times by producing its own and storing a lot of it. The family, of ample size and age gradation, was able to provide most of the skills and services for self-sufficiency by maintaining diversified production and well-knit social organization. This competence and unity were maintained long after the necessity had disappeared. As time is measured in American history, the life of this society, and its vitality, was extraordinary.

Looking back from the ease of the present, these elder days may seem to have been a time of lonely and hard isolation. It was only toward the end of the period that the telephone and rural mail delivery were added. The prairie lacked wet weather roads. In the hill sections, ridge roads might be passable at most times; on the plains, winter was likely to be the season of easiest travel, spring that of immobilization by mud. The country doctor was expected to, and did, rise above any emergency of weather. Life was so arranged that one did not need to go into town at any particular time. When the weather was bad the activites of the family took place indoors or about the farmyard. In our retrospect of the family farm as it was, we may incline to overstress its isolation. The American farmstead did not have the sociability of rural villages of Europe or of Latin America, but the entire family had duties to learn and perform and times of rest and diversion. It depended on a work morale and competence, in which all participated and in which its members found satisfaction. Perhaps it suffered less social tensions and disruptions than any other part of our society.

Though living dispersed, the farm families were part of a larger community which might be a contiguous neighborhood or one of wider association. The community in some cases got started on the Boone pattern of a settlement of kith and kin. A sense of belonging together was present to begin with or it soon developed. The start may have been as a closed community; it was likely to continue in gradual admission of others by some manner of acceptance. Consanguinity, common customs, faith or speech were such bonds that formed and maintained viable communities through good times and bad. The Mennonite colonies are outstanding examples. The absence of such qualities of co-option is shown in the Cherokee strip, opened as a random aggregation of strangers.

The country church played a leading part in social communication, differing again according to the particular confession. Catholic and Lutheran communicants, perhaps, had more of their social life determined by their church than did the others. Their priests and pastors were most likely to remain in one community and to exercise and merit influence on it. Parochial schools extended the social connections. Church festivals were numerous and attractive. Sunday observance was less austere. The Methodist Church, on the other hand, shifted its ministers, usually every two years. In a half century of service my grandfather was moved through a score of charges in five states. The high periods of the Methodist year were the winter revival meetings and the camp meetings in summer after the corn was laid by. For some, these were religious experiences, for others, especially for the young people, they were sociable times, in particular the camp meeting, held in an attractive, wooded campground where one lived in cabins or tents on an extended picnic. Almost everyone belonged to some church and most found a wide range of social contacts and satisfactions thereby.

The churches also pioneered higher education, founding colleges and academies across the Middle West, from Ohio into Kansas, before the Civil War and before the Morrill Act fathered the tax-supported colleges. These church-supported small colleges, about fifty of which still exist, first afforded education in the liberal arts to the youth of the prairie states and they did so by coeducation. Their students were drawn not only from nearby but from distant places by their church affiliations. In these colleges humane learning was cultivated and disseminated. Their campuses today are the Midwest's most gracious early monuments of the civilization aspired to by its pioneers.

Country and town were interdependent, of the same way of life, and mostly of the same people. By a tradition that may go back to the town markets of Europe, Saturday was the weekday for coming to town to transact business (note the pioneer implications in the term "to trade") and to visit. The town provided the services, goods and entertainment which the farm family required. In time, it also became home for the retired farmer.

The era of the Middle Border ended with World War I. Hamlin Garland introduced the name in 1917 in his *Sons of the Middle Border*, a retrospect he made in middle age. Willa Cather, growing up on its westernmost fringe in Nebraska, drew its life in quiet appreciation in her two books written before the war, and then saw her world swept away. Some of us have lived in its Indian summer, and almost no one was aware how soon and suddenly it was to end. A quarter section was still a good size for a family farm and the farm was still engaged in provisioning itself as well as in shipping grain and livestock. It was still growing a good crop of lusty offspring. The place of the family in the community was not significantly determined by its income, nor had we heard of standard of living.

Decline of the Middle Border

The outbreak of the war in 1914 brought rapidly rising demand and prices for supplies to the Allies and to American industry. Our intervention in 1917 urged the farmer to still more production: "Food will win the War" that was to end all wars. He made more money than ever before, he had less help, he was encouraged to buy more equipment and more land. The end of the war saw a strongly industrialized country that continued to draw labor from the rural sections. Improved roads, cars, tractors and trucks made the horse unnecessary and thereby the old crop rotation broke down. Farming became less a way of life and more a highly competitive business for which the agricultural colleges trained specialists as engineers, chemists, economists to aid fewer and fewer farmers to produce more market goods, to widen their incomes against the rising cost of labor, taxes and capital needs. This became known as "freeing people from the land," so that now we have about a tenth of our population living on farms (among the lowest ratios in the world) and these are not reproducing themselves.

The Middle Border now belongs to a lost past, a past in which different ways and ends of life went on side by side. We have since defined the common welfare in terms of a society organized for directed material progress. For the present, at the least, we control the means to produce goods at will. We have not learned how to find equivalent satisfactions in jobs well done by simple means and by independent judgment that gave competence and dignity to rural work. The family farm apprenticed youth well for life there or elsewhere and it enriched the quality of American life. It will be missed.

AMERICA EMERGING

Philip Wagner

Every tribe or nation creates its own geography. Wherever human beings live there are different evaluations, embellishments, enhancements and exploitations of environment. Every people knows a proper kind of site for its houses, size for its settlements, shape for its fields, surface for its roads, sanctuary for rest and reverence. The form thus given to environment bespeaks not only economic ends but the ethical traditions, esthetic values and ecological concepts of its people.

The validity of this notion, which is the core idea of cultural geography, is easy to demonstrate in backward areas. But it has been hard to show its relevance and value, if not its theoretical validity, for a country like the United States. Cultural geographers have not been very successful in making sense of the United States as a dynamic encounter of culture and land. The best studies deal with anachronisms and peculiar isolates, apart from the modern mainstream of America, or else they seek to illuminate the past. In treating matters economic, geographers have seemed to accept the existing (or even an earlier) state of things uncritically and not to look for major change. Despite the obviously revolutionary nature of the great geographic developments taking place in the United States today, and despite the urgency of their being understood, books and articles about the geography of this country give at best a faint impression of the transformations that are taking place, and explain them not at all.

The United States is without a doubt the most interesting and chal-

[from *Landscape*, vol. 13, no. 1, Autumn 1963]

lenging of cultural geographic problems—not because it still contains
small areas of diversity but because of its very monolithic sameness and
size. If it does not foreshadow the future condition of the entire world,
it aspires to; if it does not show the total dominance of man over environment, it displays the busiest, most rapid and extensive remaking of a
landscape on a large scale that has ever taken place. And, if it does not
threaten total disaster, it is still a situation full of dangers and uncertain-
ties.

The real epic of America today is geographic. We are remaking the
soil, the water, the very air, humbling mountains and exalting plains. We
revise all distances and scales of things, and acquire a new cardinal direction—"up."

New Geographical Values

The common man knows full well what is going on and is absorbed in it.
Thinkers ponder his indifference to matters spiritual, political, moral and
esthetic, but he is far beyond them: he is at work creating a new geography that will no doubt reveal another universe of values proper to its
time.

Everyone recognizes some of the major marks of geographic change in
the United States today: freeways, low density housing, decaying central
cities, shopping centers, declining public transport, over-burdened recreation areas, industrialized agriculture, smog, water pollution, unsightly
commercial development. Although nobody can keep up with the alteration of our landscape long enough to provide a full description of it
(which would instantly be obsolete, in any case), it can be described in a
single set of terms valid anywhere in the United States.

Likewise, almost everyone can see some of the immediate reasons for
the change: the private automobile, the increase in family size, the liber-
alization of credit, the proliferation in consumer goods and the apotheo-
sis of merchandising, the deepening social and moral chasms between
generations, the development of new construction machines, materials
and processes, the abundance and adaptability of inanimate energy
sources and a prosperity built on space exploitation and armaments.

But a good description of the new American cultural landscape, or an inventory of technical innovations and social institutions, clearly does not provide the key to an understanding of our geographic drama. There is something much deeper at work, a fundamental change in our whole cultural outlook and social structure. It is implicit in the daily activity of most of us; but, despite its grand simplicity, it eludes our thought.

To begin with, the interpretation of landscape and culture must show the effects of those things which virtually all geographers study—the so-called "man-land relationships." To mean anything, as geographers know, man-land relationships must be seized in the concrete. What are these relationships?

The most obvious of them is the formal spatial one—*location.* *"Where are people?"* The more we know about where people live and work, or have lived and worked in the past, the more we can know about their actual influence on landscape.

It is important, too, for the geographer to ask how people see the land, what they expect of it. Geographic *perception* remains an almost unstudied, though crucial question; yet it is plain that concepts and perceptions inspire and also limit human efforts to modify environment.

Another relationship, probably the most vital of all, is embodied in *technique.* Human work exerts direct material effects on the environment; how are they produced? The basic dynamic relationship of man to land resides in the use of tools and techniques, and the characteristics and capacities of their operation contain the essential explanation for all environmental change. Tools and facilities of all kinds, peculiar to a given cultural realm and tradition, represent in effect a middle term between humanity and habitat. Indeed, when being used they are in themselves the most substantial of relationships between a man and the earth, and form with both a triad which can be effectively described in mechanical or para-mechanical terms.

Certain attitudes entirely within the fabric of human societies are faithfully reflected in yet another set of man-land relationships. Call them collectively *"territoriality."* Land property rights are one form of territoriality. So is national sovereignty. In simple societies, institutions of territoriality may directly control the locational and technical relationships of men to land: each cultivator labors on his own subsistence plot, each party of hunters roams its own hunting rounds. But in more complex societies, property or state power means a centralization of control, and a massing of the effects of many men's efforts in field or factory in obedience to the decisions of one. This centralization is clearly momentous for the modification of landscape.

Although of another sort, a bond of *emotion* constitutes a significant relationship of man to land. Peasants are said to love their land, and get-rich-quick exploiters to regard it lustfully and brutally. Whatever genuine emotional relationships of this sort exist may be of tremendous importance to the landscape, but though repeatedly asserted, they are difficult to capture.

I submit that the astonishing changes recently manifested in the cultural geography of the United States have come about because *all, or nearly all, of these basic man-land relationships have undergone a drastic transformation within our culture in recent years.*

Let us look briefly at these relationships in the United States, at an earlier period and today [1963]. Compare 1930 and 1960.

The change in locational relationships is, of course, glaringly apparent. In 1930 the country still exhibited a settlement pattern of dispersed rural homesteads, small service centers and fairly compact cities fringed with suburban growth, though there was already a blurring between city and country in a narrow peripheral urban zone. Farmers worked on their home lands, industrial workers in factories reasonably near their homes; commuting was relatively slight. Today, settlement in general has become less dense and more extensive. The cores of the cities are rotting and their edges trail off into vast residential tracts threaded by commercial ribbons and spotted with shopping centers. Small trading and service towns are shriveling and dying, and even farm dwellings are declining in number, giving way to a commuter-farming arrangement. People work many miles from home.

In effect, the whole scale of settlement pattern has changed. From a well-differentiated fabric of loosely scattered rural dwellings in a landscape dotted with tight urban clusters, we are shifting to a looser but more uniform spread of settlement in a relatively few large patches, separated from one another by almost a void in the remoter countryside. Residential, as against other uses of land, tends to be more sharply segregated from other uses. The distances between residence and work are accordingly magnified.

Another important feature of the change concerns occasional travel. People go farther to play. The city parks which used to be crowded on a weekend are largely deserted the year round, and the populace entertains itself tens or hundreds of miles from home. Recreational land uses have become more demanding of space, but the densities associated with them seem hardly to have decreased; more people use them.

The shift in locational relationships has the effect that areas, even very large ones, can be more precisely classified as to prevalent land use than before. A map of the United States in 1930, accurately showing recreational, agricultural, extractive, industrial, residential, institutional and commercial land use in color, would have had to be huge to be useful. But such a map today could, with a convenient scale, show quite clearly and accurately areas of some size devoted to one or another use, big enough to stand out sharply. Such maps would be especially interesting if they could depict population characteristics such as income levels and age groupings. The map of 1930 would probably distinguish better among incomes (there was more distinction) and a 1960 map would reveal a remarkable spatial segregation of "young-marrieds," "senior citizens" and other groups, to some extent overriding income differences. The latter map would also point up considerable differences in residential age-group patterns in various parts of the country, reflecting the growing importance of climate as a resource.

Changes have undoubtedly occurred in the perceptual relationship of man and land during the thirty and more years since 1930—a changing evaluation of climate as an influence on residential distribution is a case in point—but they are harder to document. For one thing, distance, with the automobile and jet airplane, is less of a limitation upon movement.

Perception of the Environment

Probably an equally fundamental modification has taken place in the average American's understanding and appreciation of his natural environment, thanks to the popularization of science and the ascendancy of universal education. Folklore has given way to a kind of folk-science; the weatherman and the nature guide are now familiar figures. Nowhere has the change been more consequential than in the introduction of the ordinary man (and wife) to the mysteries of the garden; they know all about phosphates, horticultural varieties and nematodes.

But perception of environment is still too much of a mystery for any place or period to permit of comparison of the two generations in this regard.

New techniques would appear the surest key to all of what has happened to our landscape. Tremendous energy sources like petroleum and nuclear power have been coming into their own. New ways have been found to intensify the use of soils and waters and forests.

But the real significance of technical change in America does not lie in the multiplicity of efficiency of products or processes alone; it lies far more in the new social implications built into these products. It is in the creation of new classes of users and new purposes of use for goods that the revolution has taken place. This can best be understood in conjunction with the changes in territorial relationships.

In 1930 few people owned their homes, for example, and a large percentage lived in apartments or rented houses. On the other hand, a vastly greater proportion of small industrial and commercial enterprises were owned by those who managed them. Today retail stores and service businesses and industrial plants belong to centralized corporations, and local representatives are only branches. The decisions affecting the landscape have become entrusted to far fewer individuals, most of them remote from the environmental consequences of their actions.

Thus the proportion of ownership by occupant increased in homes but declined in business establishments. Only with radically new attitudes, as well as new financial practices, could home ownership have spread to such a large part of the population. This in itself has brought

about a change in perception of environment, and in emotional relationships with it: people long for homes and get them, and (presumably)
love them.

It is notable, too, that national sovereignty has become a factor of increased importance. Governmental regulation, increased taxation (and
the free use of taxation to guide the trend of development, consciously
or unconsciously), subvention of environmental change by public agencies, tightened custodial care in public lands and waters, and the like,
express this enrichment of state territorial relationships.

In brief, there has been emerging a new complex of locational perceptual technical and territorial relationships between man and land in the
United States. But this system is not a self-generating one, nor self-ex-
planatory. It has developed, rather, as a result of the normal play of certain economic and technological forces.

The Landscape of Production

Let us assume that it is the purpose of production to make the most
efficient use of materials, machines and men to provide the highest level
of living possible for the greatest possible number of people over the largest possible area. What does this mean for the geography of production?
It means the use of only those materials that can be delivered at lowest
cost; it means the consolidation of management at a level of maximum
efficiency; it means the specialization and concentration of technical operations to achieve optimal use of all possible advantages of skill, traditions, prestige, control, resources, market and the like; finally, it means
the coordination of production with consumer demands and continuous
innovation, experiment and development. All of these considerations add
up to a narrow choice of sites and to a need for coordination among different branch sites.

These imperatives of production have led to the virtual extinction of
independent local plants and retail outlets in many fields. They have
meant the substitution of old style community leaders by the migrant,
rootless manager, deprived of full power of decision. Even more impor-

tant, they have localized raw material extraction and industrial production in relatively few, larger and more efficient units to take advantage of the most favorable conditions, and specialized assemblages of equipment and personnel in relatively minute tasks, ingeniously integrated on a national basis. Even certain kinds of residential construction, especially of large units in the urban core, now depend on giant specialized corporations, and are located in specialized zones.

This *specialization* of production is directed, however, toward the achieving of a *generalization* of consumption. As the choice of acceptable sites for a given kind of factory narrows (because of more exacting specifications) efficiency makes it possible to distribute the products over an ever wider territory and to an even larger segment of the population. A diversified and strongly regionalized pattern of consumption would probably not be compatible with a maximum level of consumption "welfare," because it would not permit large enough and specialized enough operations for maximum efficiency.

Let us turn back to the purpose of production. Leaving to one side our tremendous non-economic expenditures in space and military technology, is the rest of our production truly enlisted in the service of consumption, as it used to be understood?

Yes and no. Here is probably the most significant departure, culturally, from our earlier condition. Conventionally, economists recognize two kinds of products: producers' goods and consumers' goods. What is taking place in the United States is the creation of an intermediate kind of goods, not "consumed," but used by the individual for such production, usually non-economic, as he likes to engage in. The automobile in America is used by consumers in non-economic production. The work done in the home by the housewife, most of it with elaborate production machines, is only in slight degree economic if measured in labor-and-commodity-market terms. The home gardener, the furniture builder working in his garage, the weekend boatman and their like are engaged in expensive operations of an inherently productive character, useless in this case, but "fun." It is all a kind of *anti-production*, a dissipation of power supplies and raw materials, and an application of tools, control and time for ends that are not concerned with increasing value. The use of land for residence and the use of scarce resources as construction materials for new homes are of a piece with this tendency.

Anti-production in the Home

Anti-production, if it may be so called, demands a small, and to the economist a wasteful, scale of operations, though in the aggregate it makes very large demands on land, time and goods. On the other hand, the kind of production which lies behind it, which supplies the means for its development, calls for operations on an increasingly ambitious scale.

It was probably inevitable that productive activity sooner or later would invade the individual home, for other purposes than simple convenience. At some point, production becomes so efficient that, within the rather anachronistic framework of political territoriality, it can exceed predictable demand for consumer wares. But it must still go on producing, and what it resorts to is the kind of product that gives the individual consumer the technical and largely honorary rank of producer, while yet not admitting him into the competitive world of production for sale and profit.

Pre-mixed, do-it-yourself, ready-to-assemble, semi-products come to us in greater and greater quantity from factories and stores. Even farm produce, partly processed, tends to increase. Some people admire this uncomprehended phenomenon, others deplore it as "do-it-badly-yourself." It makes little sense, indeed, if one does not see it all against the background of changing relationships. But do these elements of our technical equipment not allow everyman to rediscover a significant relation between himself and his surroundings, to take a hand in providing for his own needs, to declare his independence in his smallest acts? Do they not free him to roam where he pleases and set up his home? Do they not give the ordinary man his own private relationship with his environment, not the passive one of former decades, but one which is dynamic and demanding?

It may be objected that the extreme, restless mobility of American families should prevent their serious involvement in a temporary environment, but this does not seem to be the case. On the contrary, this kind of involvement may be understood as a reassuring, self-asserting gesture in the face of an unwelcome threat of rootlessness. Americans are immediatists, their grasp on life is in an active present, not in the past or the future. And so, though ever moving, they everywhere improve and transform their transitory habitats, and in so doing claim them as their own.

To build, improve, landscape and remodel is *to be at home* for many of us, and these activities are all the more important if our stay is brief. Passivity toward the home environment, and inability to "express oneself" in its transformation, would make mobility for us a different and probably a dreadful and disastrous experience; it would rob us of the psychological bulwark of "home" —and in this age Americans have little else to hold to.

The placing of powerful instruments of production in the hands of the consumer does indeed enhance his independence. But it can likewise reduce his interdependence with his fellows and involve him in another order of remote, but total, thraldom to the massive industry behind it all.

There is a momentous question inscribed in the worried face of today's landscape: will our values and our way of life in future years be built around the potentialities of the individual who is so assertive with his new technical powers, or around the equally assertive centralized institutions that provide him with those powers? The great industrial tracts, the freeways, the monster apartment house projects, the giant mechanized farms, the shopping centers and the nuclear power plants introduce a new spatial scale into the landscape, in keeping with the economic and social scale of the enterprises that create them. That scale is in itself a symbol of centralized power, abstraction, uniformity and efficiency. But this tremendous complex must serve the demands of the ordinary people who require it to provide them with the means of holding it at bay, while they remain encapsulated in their semi-automated individuality amid the housing tracts. There, perhaps, are theysymbols of potential promise for a new development of freedom and creativity. But neither the homogeneous, centripetal great enterprise nor the households, heterogeneous and centrifugal, have as yet given rise to a dignified, creative and humane way of life. Those are not the values of a revolutionary period or a landscape in transition.

The cultural landscape of the United States today expresses more clearly than ever before the ambiguity of values inherent in our heritage. This conflict is part of a culture which for generations has attempted to embrace both democracy and technology. Technological advance can lead either toward a totalitarian 1984, or toward a fuller, freer, more expressive life for every one. Democracy can lead either to a fuller participation of individuals in both public affairs and private intellectual and spiritual growth, or to creeping apathy and hysterical atavism. My reading of the landscape of America today convinces me that there is hope in the fact that our productive efficiency seems directed to serve the technological liberation of the common man, and the comman man seems to take this as his birthright.

If in this landscape now emerging everyman, in confrontation with his own environment, reshaping it himself, comes to perceive anew something of his own self, and reexamines his loyalties and values and aspirations, we may experience a rebirth of the American Promise. If the great monoliths of industry and commerce and finance—and war-making— that today cast such long shadows over cities and countrysides prove to be forerunners of a cultural glaciation, the opportunity that was America is lost to mankind. For the greatest challenge of this time in the United States is not the penetration of outer space but our fuller entering into our environmental heritage.

THE NEW AMERICAN COUNTRYSIDE
AN ENGINEERED ENVIRONMENT

J. B. Jackson

By now most of us have grown used to the idea that the urban world we live in is going on changing, and not necessarily for the better. We know what is in store: more tall buildings, more vacant downtown lots, more expressways and subdivisions and neon signs. Nurtured on science fiction, World's Fair Futuramas, Sunday spreads of the visions of real estate developers, and what the French call the literature of anticipation, we recognize behind the reality which rises to obstruct our view or intensify a traffic jam the architect's or engineer's dream.

Endurance as well as courage is needed if we are to keep on riding the wave of the future; where do they come from? From many sources; among them the firm belief, passed on from generation to generation, that there is and always will be a part of the world remote from the city that we can retreat to and find ourselves. Here is where the ancient relationship between man and nature survives intact. The weekend trip, the summer vacation, the retirement years are set aside for the renewing of this bond—except that some distraction usually interferes.

The more the city expands and absorbs us, the firmer the belief in a rural paradise becomes. Our ties with the countryside no further than twenty miles from our door grow fewer; even the annual return to the family farm, a tradition still alive a generation ago, has now all but vanished. Without personal involvement we are in the dark as to what is happening on the farm—any farm. And the result is a popular image of rural

[from *Landscape*, vol. 16, no. 1, Autumn 1966]

America which bears a decreasing resemblance to reality. We see it as a pleasant, drowsy region where old-fashioned people are engaged in a kind of work less essential and less profitable with every passing year, but where life has an elemental simplicity and truth. On a more sophisticated (though no better informed) level the countryside is seen as a vast wild-life preserve resounding with birdsong, threaded by sparkling streams—ideal for recreation and something environmental designers like to label "open space." How ever we look at it, this hinterland is held to be the great antidote, spiritual as well as physical, to the evils of the city. As long as it survives unchanged we ourselves can hope to survive; urban existence is a kind of purgatory.

It so happens that the American rural landscape is composed not only of forests and lakes and mountains, but of farms and feedlots and irrigation ditches and orchards and tractor agencies and rangeland. It is a place of work, and because it *is* a place of work, hard and not always rewarding, it is at present undergoing a revolution in its way as radical as the revolution in the urban environment. Moreover this revolution is taking place entirely without help from environmental designers. No one, outside of a handful of government agencies from the Departments of Agriculture and the Interior, is trying to direct it and give it form. Thus while we keep on counting the days until we can return to the family homestead, the homestead itself has vanished and along with it much of the 19th Century landscape. Quite a different rural America is emerging, and while there are still great changes in store it is not too soon to discern its rough outline. What does it look like? How does it differ from the one we used to know and still dream about?

It can be briefly described: it has far fewer people living in it, its work is largely mechanized, and it is evolving its own attitude toward the environment.

Half a century ago there were thirty-two million Americans living on farms. Today [1966] there are less than thirteen million, and in another twenty years there will probably be not many more than ten million. An obvious result of this decline is an increase in the number of abandoned farms. In the last fifteen years more than half of these have been in the

East—particularly in the Southeast. Americans have long been familiar with the sight of deserted farmsteads on country roads: barns and houses sagging, fields choked by a second growth of trees, lanes overgrown. These have become part of our rural picturesqueness. Abandonment of this sort is on the increase; all that prevents the complete desertion of many older countrysides is the wave of exurbanites and vacationists and retired people who are willing to restore old houses in places which have either esthetic appeal or the appeal of proximity to some city.

Well, it is pleasant to have the woods back again, even though it usually means that some unlucky farmer has had to give up the fight to make a living on his own. But it is a grievous mistake to assume that every abandoned farm automatically increases the area of forest and wildland. That is an illusion common in the East, where the cult of the tree reaches almost pagan proportions. West of Missouri, roughly speaking, abandoned farmland does not revert to forest; it reverts to rangeland, "open space" or even desert, a prey to erosion and blowing soil. If the thousands of acres withdrawn and scheduled for withdrawal in the middle regions of America are ever to serve any future use—whether as wildlands or recreation tracts or even farming—they will have to have expensive care. Furthermore (to nip any arbolatry in the bud), the few trees which grow around the abandoned places in the Great Plains will die. We should not expect the abandoned farms to produce everywhere the romantic landscape of "pleasing decay," that John Piper has defined. This will remain the exclusive property of the Northeastern states; elsewhere the environmental designer will have to create it out of very scanty material.

The earliest victim of a declining farm population is the country town, and in fact the stagnating town is already typical of much of the new American landscape, especially in the High Plains. And here again we would do well not to expect any of the traditional Eastern graces. There will be few if any tree-lined, lawn-bordered streets with white mansions taking in summer guests, no handsome common surrounded by 19th Century brick buildings, still sound and trim. The ailment affecting the average small rural center (when it has no industry) is the same one affecting our cities: a moribund downtown area. In rural terms this means that Main Street is losing its vitality. The hotel is empty, many stores are empty, the depot is empty. The decline in the number of customers is one reason; another reason, which holds good in prosperous regions too, is that modern farm equipment and its servicing takes up a great deal of space, and consequently that important aspect of small town business has often moved out to the highway where there is plenty of cheap land. The chain store supermarket has followed suit, and so has the new restaurant, and the new motel has been built out there as well. If and when the environmental designer concerns himself with small town problems, this one ought to have priority.

Perhaps it can be solved in conjunction with another problem: the growth and persistence of small town slums. Across the railroad tracks, near the river with its occasional floods, out by the municipal dump, there are shanty towns and tent cities and hopelessly immobile mobile homes where a racial minority group lives. This is all that is left of a much larger group which once supported itself after a fashion as stoop labor on the surrounding farms. Mechanization has thrown them out of work for good; but they stay on for lack of a better place to go, living more and more miserably. It is ironical that small town slums are typical of the most prosperous (because most highly mechanized) areas. The rich Delta cotton lands of Mississippi boast the highest percentage of unemployment in the South: more than 20%. This new feature of the American countryside threatens to proliferate thoughout the South, the irrigated Southwest and California.

It will undoubtedly take us time to get used to these and other indications of a population decline; we think of America as forever booming and expanding. It is true we are no longer disturbed by the abandoned one room school or the crossroads General Merchandise; but how will we take the abandoned, more or less modern, high school with monster gymnasium? The abandoned drive-in movie with rows of empty stanchions emerging from the weeds, the abandoned shopping center? We will see them, not only in North Dakota but in Texas and Florida and Kansas and elsewhere.

Fewer people, fewer workers brings on mechanization, or is it mechanization which means fewer people because of fewer jobs? Whichever is the case, mechanization of farmwork is the most conspicuous hallmark of the new rural landscape. This dawns on us long before we reach the farms themselves. I mentioned that the farm equipment dealers have all moved out to the more spacious highway strip. Actually this is no new development, but in the years to come, it is safe to say, it will assume spectacular proportions. We will have to develop an appreciative eye for this double row of immense, gleaming, bright-colored machines—tractors, combines, harvesters, pickers of all sorts, bulldozers, landplanes, wheel scrapers, self-propelled irrigation and spraying systems, balers, trailers, trucks, not to mention stacks of aluminum pipe and gas storage tanks—all of them more magnificent than their current versions—and more expensive. Some of them will be for sale, but more and more of them will be for rent, or owned and operated by what are called package farming outfits. California already has such enterprises which contract with farmers to do the fumigating and fertilizing of the soil, the preparing of the seed bed and the planting of the seed, the insecticide spraying and finally the harvesting. For tomatoes the service is $95 an acre. Since the cost of equipment is steadily increasing, and the techniques of farming are becoming steadily more complex, services of this sort are bound to become popular.

The rural highway strip lined with new farm machinery is already an impressive spectacle. It would be even more impressive and more efficient if it were properly and imaginatively laid out. This is the sort of improvement which the highway designer or the landscape architect is quite capable of undertaking. The urbanist problems of small towns are on a small scale, perhaps, but agricultural mechanization, potentially the source of local employment in servicing and repairs, has brought serious traffic and parking hardships with it.

Farther out along the strip, beyond the motels and the giant truck stop and the last rusty used car lot, lies the local landing field. Once a neglected facility, used chiefly by city sportsmen and a few well-to-do amateur flyers, it has become a vital element in the new landscape. There are planes here for spraying and dusting and seeding; in some places there are helicopters whose function it is to hover over fruit trees and shake down the ripe fruit. And it is only a matter of time before we see government planes for what is now called "remote sensing." By means of infrared photography it is possible to check from the air the condition of crops and forest and rangeland, and spot plant diseases long before they are visible to the naked eye, especially in the vast fields which cannot be inspected from the ground.

What about the appearance of the farms themselves? That depends on where they are. It should go without saying that a Northeastern dairy landscape will not resemble a Montana wheat landscape or an Arizona citrus landscape or an Alabama cotton landscape in 1986 any more than it does in 1966. But it is possible to predict that all farms in the future—commercial farms at least—will have certain traits in common, traits which are not yet widespread.

They will (for one thing) be larger. One result of a diminishing farm population is that many small holdings are consolidated into a few size-able ones, though not necessarily by outright purchase. Averages mean very little in a country as diversified as ours; just the same it is worth noting that whereas only twenty years ago the average holding was 190 acres it is now [1966] 357 acres. The increase has actually been proportionally much greater in the Midwest and the Far West. Generally speaking, the smaller the farm the greater the probability of its being either abandoned or absorbed. In one relatively prosperous Illinois county, fifty farmers are going out of business each year. Most of them were working 100 acres or less, and their land has been quickly taken up by larger operators. One thing which has contributed to this expansion is the increased mobility of farm machinery. With faster moving tractors, farmers no longer hesitate to acquire land several miles away from their headquarters. "Tractors with Lugs Prohibited" is a once common highway sign you no longer see; that is because they now have pneumatic tires.

A landscape composed of large farms or farm complexes will mean a considerable change in many ways: fewer fences and hedgerows, less variety in the vegetation; more abandoned barns and out-of-date equipment in the midst of intensively cultivated fields; bales of alfalfa bulging out of deserted farm house windows. If we were to suddenly come upon this new landscape for the first time in our lives, our reactions would be not unlike Gulliver's when he first glimpsed Brobdingnag: marveling at the immense fields, the great rows of trees, the giant stands of grass and wheat; but also noting the coarseness of detail: the lack of men, of animals, of small woodlots, of isolated barns and sheds.

Nevertheless what we are more likely to notice is the artificial topography of the landscape, the manner in which the cultivated land has been remodeled.

Every American farmer in the past has sought to change the topography and vegetation of his land to suit his needs: he has cut down stands of trees, dammed and regulated streams, drained marshes, cleared the fields of stones and stumps. But his intent was usually to "assist" nature, to encourage its more productive aspects. Whether this reflected a kind of piety or merely a lack of ability to do more in the way of modification is a point which could be usefully debated; but in former times the natural order, slightly altered by man, was held to be the basis of successful agriculture. To express this in contemporary language, the traditional farmer "studied natural systems and focused his attention on discovering and applying natural laws to the behavior of these systems and on explaining the relationships and interaction of separate parts."

The kind of modification which the modern farmer undertakes is entirely different; different not only in scale but in purpose; for its purpose is to create an entirely new and artificial setting for his work. The ultimate aim is a man-made topography, a man-made soil, a man-made crop, all part of a new production process.

More and more this is what is being achieved. The development of powerful and versatile earth moving equipment both during the war and in subsequent highway and reclamation programs has enabled any farmer who can afford it to remake his farm. To the right-thinking suburbanite the bulldozer is the very embodiment of ruthless destruction. To the irrigation farmer or the farmer trying to consolidate a variety of uneconomical holdings or the farmer cramped for space—and they exist—the bulldozer is a godsend. Thanks to it he can level land to be put under the ditch, suppress washes and gullies, root out unwanted vegetation, terrace slopes which are too steep for profitable cultivation. Furthermore, earth moving equipment can regenerate the soil by means of deep plowing, it can conserve moisture and prevent erosion. The danger of abusing this tremendous power is easy to illustrate, so much so that the layman is likely to forget its beneficial capacity. The farmer is no longer strictly confined to one type of terrain, or one area. Within reasonable limits he is free to exploit good soil wherever he finds it. So a little less vehemence, please, about urban sprawl gobbling up all the farm land—particularly when the vehemence is not entirely sincere.

The chief value of land remodeling is however neither soil and water conservation nor the creation of more cultivable land: it is the creation of large, flat, uniform surfaces for modern farm equipment. The larger the farm the more economic justification for mechanization; this is the best—indeed, the only—way of saving time and labor. The tractor is perhaps the basic piece of mechanical field equipment; there are about five million of them in the United States. It is a capable and willing servant, but it is inclined to be exacting. Once it crept about the fields at 5 m.p.h.; now it works at three times that rate. But in return it requires large fields of uniform level and texture, rectangular in layout. Thus in addition to encouraging larger farms, mechanical equipment encourages an increasingly artificial topography. Nor is this the last of its demands; as it begins to take over more and more of the work once done by hired help—the planting of the seeds, weed control, fertilizing, cultivating and harvesting, it insists on further changes. The human hand, however much it may cost per hour or per bushel, knows how to cope with unexpected irregularities: unevenness in the ground, plants not ripened or out of alinement. The harvesting machine is less tolerant and less adaptable: the seed has to be accurately planted and so spaced that the machine can deal with it. So with the advent of the newer, more complex processing machines there come into the landscape new kinds of rows, new spaces and intervals, and new locations for roads and irrigation apparatus.

The latest demand of the machine is the most revolutionary of all: it wants nothing less than new varieties of plants that are more uniform, easier to pick and less perishable. The demand is being met with great dispatch; the University of California at Davis, concerned by the abrupt end of migratory labor, is undertaking to develop some twenty new varieties of commercially produced fruits and vegetables for harvesting machines.

The assembled experts on soils, agricultural engineering, plant pathology, genetics and dietetics hope to come up within the next two years with more amenable canteloupes, lemon trees which grow to a prescribed height, grape bunches and strawberries with longer, more convenient stems. These varieties will in turn require new field layouts for irrigation and access, and new fertilizing and weed killing techniques.

Once considered an unchanging element, the soil is now subject to drastic modification. The chemical fertilizers, which have only begun to come into their own, have made more intensive cultivation possible, and rotation of crops is much less significant in the maintenance of fertility. The ultimate in soil manipulation is probably the procedure followed in the Monrovia Nursery in California. All topsoil has been removed from the 250 acre tract, deposited in piles and chemically treated, then put in cans. Completely terraced and scraped, the land is merely a platform on which to set the plants and operate the nursery. To many commercial farmers the soil has become one item in the production process—as much subject to improvement as is the machine or the crop itself.

In a much less familiar guise the machine dominates another part of the farm: the headquarters. Swept clean of all the usual farmyard clutter of broken and obsolete equipment, chickens, haystacks and manure, the area looks much like an industrial plant, with its long, metal sheds and barns and its neatly parked farm equipment. But the machines which actually dominate the headquarters are not those on wheels; they are the working structures themselves.

The barn is not a shelter any more, it is a machine closely involved in the productive process. If we are prepared to accept that interpretation then we are well on the way to understanding the nature of the revolution in the rural landscape. For much of this revolution is a matter of new definitions. A current writer on farm engineering makes two important points. "The escape of the tractor from its identity with the horse occurred when its design and use were related to the inherent utility it offered and not to replacement or substitution. . . . Structures today are evolving as units that contribute to the dynamics of farming. Many will become increasingly difficult to classify uniquely as structures or ma-

chines, and their function will range far beyond housing and storage."
Viewed in this light, many new barns are machines which process their
contents: provide animals with the correct amount of light and heat and
air and space; preserve, prepare and distribute feed. The best of them are
designed by animal physiologists working with engineers and systems experts. In the future they and their specialized equipment will be produced
in factories and assembled on the farm; no doubt they will be eventually
turned in toward a newer model. What seems to be evolving in our new
rural landscape is a form and concept of utilitarian architecture which
the city as yet knows little about.

The structure ceases to be a shelter or a container and becomes the totally engineered artificial environment; the term could also be applied to
the cultivated field. "The engineer and the farmer," the *Farm Quarterly*
wrote, "have joined forces to control more completely the environment
in which plants and animals are produced. The nutrient needs of animals
and plants, the moisture requirements of plants and how to control them,
have become valuable facets of farming. In the world of plants the race
has been to create through landforming, irrigation techniques, knowledge
of weather patterns, the use of nutrients, spacing and environment which
will permit each plant to approach its maximum potential. So too do animals respond to environment. Buildings are designed to allow concentration of animals, leading to increased production volume with less labor.
. . . While facilities in many cases are designed to the requirements of
plants and animals, in others the plant or animal is adapted to the facility." And the trend seems clearly to be in the latter direction: the modification of the animal or plant to suit the engineered environment. A
professor in a Midwestern agricultural college predicts that we will ultimately raise all our livestock in buildings, just the way we do chickens.
When that day comes there will be no more fences on the landscape.

Does all this sound like an up-to-date version of the factory in the
fields? Is the landscape we have been describing really anything more
than the impress of new food processing and distributing methods on agriculture, an essentially alien element in the rural scene? It is impossible
to deny that the contemporary farmer is very much part of the world of

chain stores, credit, government controls, technology. The hideous word "agribusiness" enjoys a wide popularity as the farmer seeks to identify himself with this other community, and he often seems proud of his new economic status; nothing irritates him more than wistful references to farming as a way of life. In current agricultural literature there is a notable absence of any cult of agrarianism, any reference to those beliefs which only a century ago were so widely and proudly held: that the farmer because of his hardy independence, his closeness to nature, his love of the land, was a superior type of American. Now the overriding theme is always the way the farmer can make both ends meet and plan for expansion. "How I Turned my Old Brood Sows into Dollars" —variations on this theme recur in every issue of every farm journal.

It is easy, therefore, for the outsider to pass severe judgment on the modern landscape and the men who are bringing it into being; the farmer has unwittingly created a very unattractive image of himself. It is likewise easy to dismiss this landscape as a violent modification of the natural environment entirely for the sake of more money, and with no thought of the long range consequences. If this were indeed the basic motive of the American farmer we would have good cause to fear for the future. Despite vastly increased agricultural yields and greatly improved techniques, there is a widespread conviction that the farmer's faith in chemical fertilizers, insecticides, growth stimulants, genetic experimentation will eventually lead to disaster. And such easily observable practices as irrigation in regions of adequate rainfall, clean farming, continuous cultivation of one crop, increasingly narrow rows, convince casual observers that farmers are inspired entirely by eagerness for profits.

But even if all these procedures fail we are not likely to see a return to traditional methods. And that is because the farmer, despite his bad public relations, is not entirely a businessman. He still is, and always will be, a designer of environments. His forebears were the same: they sought to design them according to what they conceived to be an order prescribed by nature or divine law; as a product of this century the modern farmer is designing by means of constant experimentation. If present techniques backfire he will not hesitate to drop them in favor of others. Dollars are what he is after, of course; but he is also after something like an insight into the truth.

Meanwhile he appears to be formulating a highly intellectual, impersonal man-environment relationship which repels many because it is so at variance with the nature philosophy formulated by two centuries of urban culture. That is why it is dangerous to assume that the American countryside can continue to play its traditional role in the lives of city dwellers. When we take a second look we are bound to recognize that the rural landscape now coming into existence has little to offer us in the way of pleasure or recreation.

It has nothing to offer the lover of romantic beauty or the seeker after undisturbed nature; it has nothing in the way of pleasure to offer the inhabitant of small country towns or for that matter the farmer himself. Yet all of these need what the country used to provide. There are two solutions: one, advocated by the beautificationists, calls for the introducing of esthetic amenities throughout the landscape, a cosmetic treatment to disguise or at least adorn the workaday features of the country (tree planting, as might be expected, plays an important role). The other solution is to design and create on a large scale the appropriate settings for recreation and locate them near the engineered environment but not in it. It is essential that the two be kept distinct and separate; they should not overlap or blend.

There is plenty of room in our countryside for both kinds of landscape—the farmer's landscape of work, the city dweller's landscape of play, and there is plenty of talent among environmental designers to produce them. So far we have been singularly timid in America in the creating of artificial landscapes for recreation: the so-called English park is about as far as we have dared to go. But there is one lesson we can learn from the modern farmer: how comparatively easy it is to engineer the environment. There are few technological limits to our capacity to transform the land to suit our needs; all that is lacking is imagination and a sense of purpose.

BOUNDARIES AND TERRITORIES

Front yards are a national institution—essential to every home, like a
Bible somewhere in the house. —J. B. Jackson

. . . a new human landscape is beginning to emerge in America. . . . That
landscape however, is not yet here. In the early dawn where we are we
can perhaps discern its rough outlines, but we cannot have any real feeling for it. We cannot possibly love the new, and we have ceased to love
the old. The only fraction of the earth for which an American can still
feel the traditional kinship is that patch of trees and grass and hedge he
calls his yard. —J. B. Jackson

. . . there seems to have come a time in this country when all boundaries,
all traditional divisions of space, whatever their original purpose, were
seen not as desirable forms of protection but as obstacles—obstacles to
beauty, to efficiency, to sociability, to change. —J. B. Jackson

GHOSTS AT THE DOOR

J. B. Jackson

The house stands by itself, lost somewhere in the enormous plain. Next to it is a windmill, to the rear a scattering of barns and shelters and sheds. In every direction range and empty field reach to a horizon unbroken by a hill or the roof of another dwelling or even a tree. The wind blows incessantly; it raises a spiral of dust in the corral. The sun beats down on the house day after day. Straight as a die the road stretches out of sight between a perspective of fence and light poles. The only sound is the clangor of the windmill, the only movement the wind brushing over the grass and wheat, and the afternoon thunderheads boiling up in the western sky.

But in front of the house on the side facing the road there is a small patch of ground surrounded by a fence and a hedge. Here grow a dozen or more small trees—Chinese elms, much whipped and tattered by the prevailing gale. Under them is a short expanse of bright green lawn.

Trees, lawn, hedge and flowers—these things, together with much care and a great expenditure of precious water, all go to make up what we call the front yard. Not only here on the Western farmstead, but on every one of a million farms from California to Maine. All front yards in America are much the same, as if they had been copied from one another, or from a remote prototype.

They are so much part of what is called the American Scene that you are not likely to wonder why they exist. Particularly when you see them in the East and Midwest; there they merge into the woodland landscape and into the tidy main street of a village as if they all belonged together. But when you travel west you begin to mark the contrast between the yard and its surroundings. It occurs to you that the yard is sometimes a very artificial thing, the product of much work and thought and care. Whoever tends them so well out here on the lonely flats (you say to yourself) must think them very important.

And so they are. Front yards are a national institution—essential to every home, like a Bible somewhere in the house. It is not their size which makes them so. They are usually so small that from a vertical or horizontal distance of more than a mile they can hardly be seen. Nor are they always remarkable for what they contain. No; but they are pleasant oases of freshness and moving shade in the heat of the monotonous plain. They are cool in the summer and in the winter their hedges and trees do much to break the violence of the weather. The way they moderate the climate justifies their existence.

They serve a social purpose, too. By common consent the appearance of a front yard, its neatness and luxuriance, is an index of the taste and enterprise of the family who owns it. Weeds and dead limbs are a disgrace, and the man who rakes and waters and clips after work is usually held to be a good citizen.

So this infinitesimal patch of land, only a few hundred square feet, meets two very useful ends: it provides a place for outdoor enjoyment, and it indicates social standing. But in reality does it always do those things?

Many front yards, and by no means the least attractive, flourish on the Western ranches and homesteads many miles from neighbors. They waste their sweetness on the desert air. As for any front yard being used for recreation, this seems to be a sort of national myth. Perhaps on Sunday afternoons when friends come out from town to pay a visit chairs are tentatively placed on the fresh cut grass. For the rest of the week the yard is out of bounds, just as the now obsolete front parlor always used to be. The family is content to sit on the porch when it wants fresh air. It admires the smooth lawn from a distance.

The true reason why every American house has to have a front yard is probably very simple: it exists to satisfy a love of beauty. Not every beauty, but beauty of a special, familiar kind; one that every American can recognize and enjoy, and even after a fashion recreate for himself.

The front yard, then, is an attempt to reproduce next to the house a certain familiar or traditional setting. In essence the front yard is a landscape in miniature. It is not a garden; its value is by no means purely esthetic. It is an enclosed space which contains a garden among other things. The patch of grass and Chinese elms and privet stands for something far larger and richer and more beautiful. It is a much reduced version, as if seen through the wrong end of a pair of field glasses, of a spacious countryside of woods and hedgerows and meadow.

Such was the countryside of our remoter forebears; such was the original, the proto-landscape which we continue to remember and cherish, even though for each generation the image becomes fainter and harder to recall.

Loyalty to a traditional idea of how the world should look is something which we not always take into account when analyzing ourselves or others. Yet it is no more improbable than loyalty to traditional social or economic ideas or to traditional ideas in art. The very fact that we are almost completely unaware of our loyalty to a proto-landscape allows us to express that loyalty with freedom. We have not yet been made ashamed of being old-fashioned. But what precisely is that landscape which our memory keeps alive and which an atavistic instinct tries to recreate?

It is not exclusively American. It is not New England or Colonial Virginia or Ohio, it is nothing based on pictures and vacation trips to the East. It is northwestern Europe. Whatever the ethnic origin of the individual American, however long his family may have lived in this country, we are all descendants, spiritually speaking, of the peoples of Great Britain and Ireland, of the Low Countries, and to a lesser extent of northern France and western Germany. It was from those countries that the colonists transferred the pattern of living which is still the accepted pattern of living in North America. It may not remain so much longer, but that is something else again. We are all of us exiles from a landscape of streams and hills and forests. We come from a climate of cold dark winters, a few weeks of exuberant spring, and abundant snow and rain. Our inherited literary and popular culture both reflect that far-off environment, and until recently our economy and society reflected it too.

For almost a thousand years after the collapse of the Roman Empire the history of Europe was the history of a slow and persistent de-forestation. When the Classic civilization began to die, Europe ceased to be one unit and became two. The region around the Mediterranean preserved a good deal of the Roman heritage; for the most part its population did not greatly change; and the land remained under cultivation. But for several reasons the entire northwestern portion of the Empire—Great Britain, the Lowlands, northern France and western Germany began to revert to wilderness. Roads, towns, cities and farms were gradually abandoned, fell into ruin, and in time were hidden by brush and forest. The peoples whom we call the Barbarians and who later moved in from the East had thus to reclaim the land all over again. They were obliged to take back from the forest by main force whatever land they needed for farms and pastures and villages. They were pioneers no less tough than those who settled our own West. Their numbers were so few and their means so primitive that every lengthy war and every epidemic saw much newly cleared land revert to undergrowth once more. It was not until a century ago that the last wastelands on the continent were put under cultivation. The whole undertaking was an extraordinary phase of European history, one which we know very little about. How well it succeeded is shown by the fact that Holland, now a land of gardens originally meant "Land of Forests."

Could this incessant warfare with the forest fail to have an effect on the men who engaged in it? Does it not help to explain an attitude toward nature quite unlike that of the peoples farther south? The constant struggle against cold and solitude and darkness, the omnipresent threat of the wilderness and the animals that lived in it in time produced a conviction that there was no existing on equal terms with nature. Nature had to be subdued, and in order to subdue her men had to study her and know her strength. We have inherited this philosophy, it sometimes seems, in its entirety: this determination to know every one of nature's secrets and to establish complete mastery over her; to love in order to possess and eventually destroy. It is not a point of view which has worked very well here in the West. If we had thought more in terms of cooperation with a reluctant and sensitive environment, as the Mediterranean people still do, and less in terms of "harnessing" and "taming," we would have not made such a shambles of the Southwestern landscape.

That aggressive attitude is however only part of what the earliest farmers in northern Europe bequeathed us. Since they created the human landscape themselves and under great difficulties, they had a deep affection for it. They looked upon the combination of farmland and meadow and forest as the direct expression of their way of life. It was a harsh and primitive landscape, just as by all accounts it was a harsh and primitive way of life, but it was not lacking in a sentiment for the surrounding world, nor an element of poetry. The perpetual challenge of the forest stirred the imagination as did no other feature in the environment. It was the forest where the outlaw went to hide, it was there that adventurous men went to make a new farm and a new and freer life. It teemed with wolves, boars, bears and wild oxen. It contained in its depths the abandoned clearings and crumbling ruins of an earlier civilization. It was a place of terror to the farmer and at the same time a place of refuge. He was obliged to enter it for wood and game and in search of pasture. For hundreds of years the forest determined the spread of population and represented the largest source of raw materials; it was an outlet for every energy. Its dangers as well as its wealth became part of the daily existence of every man and woman.

When at last it was removed from the landscape our whole culture began to change and even to disintegrate. A Frenchman has written a book to prove that the decline in popular beliefs and traditions (and in popular attitudes toward art and work and society) in his country was the direct outcome of the destruction a century ago of the last areas of untouched woodland. If he is correct, how many of those traditions can be left among us who have denuded half a continent in less than six generations? The urge to cut down trees is stronger than ever. The slightest excuse is enough for us to strip an entire countryside. And yet—there is the front yard with its tenderly cared-for Chinese elms, the picnic ground in the shadow of the pines, and a mass of poems and pictures and songs about trees. A Mediterranean would find this sentimentality hard to understand.

The old ambivalence persists. But the reverence for the forest is no longer universal. Our household economy is largely free from dependence on the resources of the nearby forest, and any feeling for the forest itself is a survival from childhood associations. Until the last generation it might have been said that much of every American (and northern European) childhood was passed in the landscape of traditional forest legends. Time had transformed the reality of the wilderness into myth. The forest outlaw became Robin Hood, the vine-grown ruins became the castle of Sleeping Beauty. The frightened farmer, armed with an ax for cutting firewood, was the hero of Little Red Riding Hood and the father of Hänsl and Gretl. In a sense, our youngest years were a re-enactment of the formative period of our culture, and the magic of the forest was never entirely forgotten in adult life. Magic, of course, is part of every childhood; yet if a generation grew up on the magic of Superman and Mickey Mouse and Space Cadet instead, if it lived in the empty and inanimate landscape which provides a background for those figures, how long would it continue to feel the charms of the forest? How long would the Chinese elms be watered and cared for?

After the forest came the pasture, and the pasture in time became the lawn. When a Canadian today cuts down trees in order to start a farm he says he is "making land." He might with equal accuracy say that he is "making lawn," for the two words have the same origin and once had the same meaning. Our lawns are merely the civilized descendants of the Medieval pastures cleared among the trees. In the New Forest in England a "lawn" is still an open space in the woods where cattle are fed.

So the lawn has a very prosaic background, and if lawns seem to be typically northern European—the English secretly believe that there are no true lawns outside of Great Britain—that is simply because the farmers in northern Europe raised more cattle than did the farmers near the Mediterranean, and had to provide more feed.

As cattle and sheep-raising increased in importance, the new land wrested from the forest became more and more essential to the farmer; he set the highest value on it. But to recognize the economic worth of a piece of land is one thing; to find beauty in it is quite another. Wheat fields and turnip patches were vital to the European peasant, yet he never, as it were, domesticated them. The lawn was different. It was not only part and parcel of a pastoral economy, it was also part of the farmer's leisure. It was the place for sociability and play; and that is why it was and still is looked upon with affection.

The common grazing land of every village is actually what we mean when we speak of the village common, and it was on the common that most of our favorite group pastimes came into being. Maypole and Morris dances never got a foothold in northern America and for that we can thank the Puritans. But baseball, like cricket in England, originated on the green. Before cricket the national sport was archery, likewise a product of the common. Rugby, and its American variation football, are both products of the same pastoral landscape, and golf is the product of the very special pastoral landscape of lowland Scotland. Would it not be possible to establish a bond between national sports and the type of terrain where they developed? Bowling is favored in Holland and near the Mediterranean—both regions of gardens and garden paths. A continental hunt is still a forest hunt; the English or Irish hunt needs a landscape of open fields and hedgerows. Among the many ways in which men exploit the environment and establish an emotional bond with it we should not forget sports and games. And the absence among certain peoples of games inspired by the environment is probably no less significant.

In the course of time the private dwelling took over the lawn. With the exclusion of the general public a new set of pastimes was devised: croquet, lawn tennis, badminton, and the lawn party. But all of these games and gatherings, whether taking place on the common or on someone's enclosed lawn, were by way of being schools where certain standards of conduct, and even certain standards of dress were formed. And in an indefinable way the lawn is still the background for conventionally correct behavior. The poor sport walks off the field; the poor citizen neglects his lawn.

Just as the early forest determined our poetry and legend, that original pasture land, redeemed from the forest for the delectation of cows and sheep, has indirectly determined many of our social attitudes. Both are essential elements of the proto-landscape. But in America the lawn is more than essential; it is the very heart and soul of the entire front yard. We may say what we like about the futility of these areas of bright green grass; we may lament the waste of labor and water they represent here in the semi-arid West. Yet to condemn them or justify them on utilitarian or esthetic grounds is to miss the point entirely. The lawn with its vague but nonetheless real social connotations is precisely that landscape element which every American values most. Unconsciously he identifies it with every group event in his life: childhood games, commencement and graduation with white flannels or cap and gown, wedding receptions, "having company," the high school drill field and the Big Game of the season. Even the cemetery is now landscaped as a lawn to provide an appropriate background for the ultimate social event. How can a citizen be loyal to that tradition without creating and taking care of a lawn of his own? Whoever supposes that Americans are not willing to sacrifice time and money in order to keep a heritage alive regardless of its practical value had better count the number of sweating and panting men and women and children, pushing lawn-mowers on a summer's day. It is quite possible that the lawn will go out of fashion. But if it does it will not be because the toiling masses behind the lawn-mower have rebelled. It will be because the feeling of being hedged in by conventional standards of behavior has become objectionable.

To hedge in, to fence in; the language seems to shift in meaning and emphasis almost while we use it. Until not long ago neither of those words meant to keep in. They meant to keep out. A fence was a de-fense against trespassers and wild animals. The hedge was a coveted symbol of independence and privacy. Coveted, because it was not every farmer who could have one around his land.

Like the lawn and the tree, the hedge is something inherited from an ancient agricultural system and an ancient way of life. The farming of the Middle Ages is usually called the open-field system. Briefly, it was based

on community ownership (or community control) of all the land—ownership by a noble amounted to the same thing—with fields apportioned to
the individual under certain strict conditions. Among them were rules as
to when the land was to lie fallow, what day it was to be plowed, and
when the village cattle were to be allowed to graze on it. Much modified
by social and economical revolutions, the openfield system still prevails
over much of northern Europe. Fences and hedges, as indications of
property lines, naturally had no place in such a scheme.

In the course of generations a more individualistic order came into
being, and when for several good reasons it was no longer desirable to
have the cattle roaming at will over the countryside the first thing to appear, the first change in the landscape, was the hedge. With that hedge to
protect his land against intruders of every kind the individual peasant or
farmer began for the first time to come into his own, and to feel identi-
fied with a particular piece of land. He did not necessarily own it; more
often than not he was a tenant. But at least he could operate it as he saw
fit, and he could keep out strangers.

Each field and each farm was defined by this impenetrable barrier. It
served to provide firewood, now that the forests were gone, shelter for
the livestock, and a nesting place for small game. Most important of all
the hedge or fence served as a visible sign that the land was owned by one
particular man and not by a group or community. In America we are so
accustomed to the fence that we cannot realize how eloquent a symbol
it is in other parts of the world. The Communist governments of Europe
do realize it, and when they collectivize the farms they first of all destroy
the hedgerows—even when the fields are not to be altered in size.

The free men who first colonized North America were careful to bring
the hedge and fence with them, not only to exclude the animals of the
forest, but as indications of the farmers' independent status. Hedges and
fences used to be much more common in the United States than they are
now. One traveller in Revolutionary New England enumerated five different kinds—ranging from stone walls to rows of up-ended tree stumps.
In Pennsylvania at the same period fields were often bordered with privet.
As new farms were settled in the Midwest every field as a matter of
course had its stone wall or hedge or privet or hawthorn, or permanent

wooden fence. And along these walls and fences a small wilderness of brush and vine and trees soon grew, so that every field had its border of shade and movement, and its own wild life refuge. The practice, however inspired, did much to make the older parts of the nation varied and beautiful, and we have come to identify fences and hedges with the American rural landscape at its most charming.

As a matter of fact the hedge and wooden fence started to go out of style a good hundred years ago. Mechanized farming, which started then, found the old fields much too small. A threshing machine pulled by several teams of horses had trouble negotiating a ten acre field, and much good land was wasted in the corners. So the solution was to throw two or more fields together. Then agricultural experts warned the farmers that the hedge and fence rows, in addition to occupying too much land, harbored noxious animals and birds and insects. When a farm was being frequently reorganized, first for one commercial crop then another, depending on the market, permanent fences were a nuisance. Finally Mr. Mr. Glidden invented barbed wire, and at that the last hedgerows began to fall in earnest.

There were thus good practical reasons for ridding the farm of the fences. But there was another reason too: a change in taste. The more sophisticated landscape architects in the mid-century strongly advised home-owners to do away with every fence if possible. A book on suburban gardening, published in 1870, flatly stated: "that kind of fence is best which is least seen, and best seen through." Hedges were viewed with no greater favor. "The practice of hedging one's ground so that the passer-by cannot enjoy its beauty, is one of the barbarisms of old gardening, as absurd and unchristian in our day as the walled courts and barred windows of a Spanish cloister."

Pronouncements of this sort had their effect. Describing the early resistance to the anti-fence crusade during the last century a writer on agricultural matters explained it thus: "Persons had come to feel that a fence is as much a part of any place as a walk or a wall is. It had come to be associated with the idea of home. The removal of stock was not sufficient reason for the removal of the fence. At best such a reason was only negative. The positive reason came in the development of what is really the art-idea in the outward character of the home . . . with the feeling that the breadth of setting for the house can be increased by extending the lawn to the actual highway."

Utilitarian considerations led the farmer to suppress the fences between his fields; esthetic considerations led the town and city dwellers to increase the size of their lawns. Neither consideration had any influence on those who had homesteaded the land, lived on it and who therefore clung to the traditional concept of the privacy and individualism of the home. The front yard, however, had already become old-fashioned and countrified fifty years ago; the hedge and picket fence, now thought of as merely quaint, were judged to be in the worst taste. Today, in spite of their antiquarian appeal, they are held in such disrepute that the modern architect and the modern landscapist have no use for either of them; and they are not allowed in any housing development financed by FHA.

Why? Because they disturb the uniformity of a street vista; because they introduce a dangerous note of individualistic non-conformity. Because in brief they still have something of their old meaning as symbols of self-sufficiency and independence. No qualities in twentieth-century America are more suspect than these.

It is not social pressure which has made the enclosed front yard obsolescent, or even the ukase of some housing authority, egged on by bright young city planners. We ourselves have passed the verdict. The desire to identify ourselves with the place where we live is no longer strong.

It grows weaker every year. [In 1951] one out of a hundred Americans lives in a trailer; one out of every three American farmers lives in a rented house. Too many changes have occurred for the old relationship between man and the human landscape to persist with any vigor. A few decades ago the farmer's greatest pride was his woodlot, his own private forest and the forest of his children. Electricity and piped-in or bottled gas have eliminated the need for a supply of fuel, and the groves of trees, often fragments of the virgin forest, are now being cut down and the stumps bulldozed away. The small fields have disappeared, the medium sized fields have disappeared; new procedures in feeding and fattening have caused meadows to be planted to corn, range to be planted to wheat; tractors make huge designs where cattle once grazed. A strand of charged wire, a few inches off the ground, takes the place of the fence, and can be moved to another location by one man in one day. The owner of a modern mechanized farm and even of a scientific ranch need no longer be on hand at all hours of the day and night. He can and often does commute to work from a nearby town. His children go to school and spend their leisure there, and the remote and inconvenient house on the farm is allowed to die.

All this means simply one thing: a new human landscape is beginning to emerge in America. It is even now being created by the same combination of forces that created the old one: economic necessity, technological evolution, change in social outlook and in our outlook on nature. Like the landscape of the present, this new one will in time produce its own symbols and its own beauty. The six-lane highway, the aerial perspective, the clean and spacious countryside of great distances and no detail will in a matter of centuries be invested with magic and myth.

That landscape, however, is not yet here. In the early dawn where we are we can perhaps discern its rough outlines, but we cannot have any real feeling for it. We cannot possibly love the new, and we have ceased to love the old. The only fraction of the earth for which an American can still feel the traditional kinship is that patch of trees and grass and hedge he calls his yard. Each one is a peak of a sinking world, and all of them grow smaller and fewer as the sea rises around them.

But even the poorest of them, even those which are meager and lonely and without grace, have the power to remind us of a rich common heritage. Each is a part of us, evidence of a vision of the world we have all shared.

WALLS AND FENCES

Wilbur Zelinsky

Although most laymen think of geography as having to do with the location of places or the larger design of the earth's physical features and
some of the economic activities of its human occupants, actually this
wide-ranging discipline embraces an almost endless variety of phenomena,
from the cosmic down to the sub-microscopic. Toward the latter end of
the scale, few topics are as absorbing as those ultimate units of land use:
the house lots, farmsteads, fields, and other parcels of real estate which
can be thought of as the molecules of the human geographer's universe.
To the perceptive student, their size, shape, location, arrangement, and
the means used to demarcate and enclose them often furnish surprisingly
detailed information as to the history and present functioning of a community.

European geographers have pursued such research with considerable
success for some years; but in North America this variety of microgeogra-
phy has been rather generally overlooked. Rather curiously, though,
European investigators have tended to avoid the matter of fencing: perhaps because fences are not as frequently built as in America or because
it is easier to take such commonplace objects for granted in a relatively
unchanging rural landscape. A good beginning has been made, however,
by several American students. They have ably described and classified
farm fencing and have explored the economic aspects of the subject.

My purpose here is to supplement earlier studies by contributing some
observations on the historical geography of farm fencing along the Atlan-

[from *Landscape,* vol. 8, no. 3, Spring 1959]

tic Seaboard—a germinal area of peculiar interest in that it combines early European settlement with rapid economic and cultural change during recent decades. No attempt is made to touch upon other types of enclosures, such as those used around houses, barns and other buildings. The chief sources of my information are the narratives of early visitors and several years of travel and field work on settlement features. It is only fair to state that the sum of the data so derived is not impressive and that there remain wide gaps, both in time and space, in this rudimentary account.

Very little is known of the kinds of enclosures erected by the earliest settlers of the Atlantic Seaboard. The provision of food and shelter, along with a minimum of military security, was their immediate aim. Furthermore, relatively few farms were contiguous then, so that property demarcation was not a major concern. We can agree with Mather and Hart [*Landscape,* Spring 1957] that "the stones, brush and trees that the pioneer cleared from fields could be heaped into an effective barrier against the cattle permitted to wander freely in the surrounding woods." We cannot be certain as to what fencebuilding skills had been brought from the British Isles, Germany, France, or elsewhere in Europe in the absence of adequate accounts of fencing practices in the home countries; but it is likely that under the strikingly different conditions of the New World few serious efforts were made to imitate European models. Peter Kalm, the insatiably curious Swedish naturalist, documented one of these exceptional instances in writing of the attempts by the Swedes who settled the lower Delaware Valley to build a traditional type of post-and-rail fence. After being frustrated by the lack of the proper varieties of wood and the rapid subsurface deterioration of the posts they had erected, they finally adopted other expedients.

The Sea-change in Material Culture

We can add fencing, then, to the roster of cultural items which underwent major transformations on being re-established across the Atlantic. In the case of the American language, the metamorphosis has been closely studied; but a great deal of interesting work remains to be done with such diverse topics as diet, costume, barns, place-name terminology, domestic architecture and town morphology.

After a period of simply shunting aside the debris from hastily cleared fields, the American pioneer farmer began to experiment with fence types. Writing of the farms between Boston and Albany in a volume published in 1792, Brissot de Warville listed some of the more important of these:

> These several kinds of fences are composed of different materials, which announce the different degrees of culture in the country. Some are composed of the light branches of trees; others of the trunks of trees laid one upon the other; a third sort is made of long pieces of wood, supporting each other by making angles at the end; a fourth kind is made of long pieces of hewn timbers, supported at the ends by passing into holes made in an upright post; a fifth is like the garden fences in England; the last kind is made of stones thrown together to the height of three feet.

One of the shrewdest British observers of the American scene, J. S. Buckingham, postulated a reasonable succession of enclosures in an account of his travels published in 1842. Although this particular passage deals with the countryside between Augusta and Bangor, Maine, the sequence might well apply to a much broader area:

> The usual order of succession in these, we were told, was first, the log fences, when the trunks of trees were rolled to the boundary line and formed into a wall or barrier. To this succeeded the stumps of the trees when rooted out of the earth, which requires a period of eight or ten years to make them sufficiently loose to be got out with ease; then they are taken up, and planted edgewise, with the roots outward to the road and the trunk-parts inward to the fields. These, in their turn, get displaced by the more perfect paling fence, or the stone-wall; and all these are so many regularly advancing steps in the settlement and improvement of a dis-

trict, like the log hut, the frame house, the brick dwelling, and the stone mansions.

The stump fence attracted attention because of its unusual appearance
and its complete effectiveness as a barrier. Harriet Martineau in 1837
called it "a singular fence, which would be perfect but for the expense of
labour required . . . needing no mending, and lasting the 'for ever' of this
world." And, indeed, the extreme durability of this fence is attested to
by the survival of some examples, possibly more than a century old, in
certain out-of-the-way corners of the upper Middle West. But although
there is an intriguing reference by one English traveler (W. N. Blane in
1824) to "an ingenious machine . . . by which these stumps are torn up
by the roots and converted into an excellent fence," the stump fence was
too difficult, expensive and space-consuming to achieve wide popularity.

The various types of rail and board fences which seem to have attained
general acceptance at fairly early dates are probably all of American origin. The most elementary—the worm or zigzag fence—was being built at
least as early as 1685 in Massachusetts and has had an Indian origin as-
cribed to it by Raup and Leechman, but this remains a moot point. Certain kinds of rail fences have been built in Scandinavia and the Alpine
regions of Europe, but it is difficult to see how there could have been
any direct connection between these and the American forms. The worm
and somewhat more advanced stake-and-rider fences, both of which call
for a maximum of lumber and a minimum of labor, occurred throughout
the entire range of the American frontier, while the post-and-rail and
board fences, which demand greater craftsmanship, were restricted to
areas of more intensive cultivation. The stone fence, which also requires a
good deal of hard labor, was an expedient adopted only in the northern
half of the Atlantic Seaboard; and the hedge, a popular device in England
and various regions on the Continent, was adopted only infrequently.

Travelers' Reports

We are fortunate in having detailed descriptions of the American agricultural landscape toward the end of the Colonial period from the pens of

two keen observers: Peter Kalm, who traversed the area between Maryland and Quebec in his journeys from 1748 to 1751, and the anonymous author of *American Husbandry* (published in London in 1775), probably either Dr. John Mitchell, the Anglo-American botanist and physician, or Arthur Young, the prolific British author on agricultural subjects. In describing the countryside near Philadelphia, Kalm stated that the post-and-rail fence was dominant, but also notes the presence of plank fences and privet hedges. It is interesting to find that the dearth of suitable fencing material in this well-settled area was already such that chestnut and red cedar were being shipped in from distant points. Kalm gives the impression that the worm fence was dominant in New York, New Jersey, and the more recently occupied parts of Pennsylvania; but he also writes of the environs of Albany that "the fences were made of pine boards of which there is an abundance in the extensive woods." The fences built by the French farmers of Quebec are said to resemble closely the post-and-rail fences common in Sweden, a fact which we may attribute to coincidence.

The author of *American Husbandry* offers a description of Pennsylvania fences which corresponds closely with Kalm's. Of Virginia and Maryland, he says simply that "their fences are extremely incomplete, and kept in very bad order," while the passages on South Carolina and Georgia indicate that fences were as yet of little importance. The situation in North Carolina was most primitive with "the woods . . . all in common, and people's property has no other boundary or distinction than marks cut in trees, so that the cattle have an unbounded range."

For the next century we must rely on travel accounts left by the many European visitors who toured North America. References to fencing in the Southern states are rare, but such glimpses as we have indicate the persistence of unfenced fields and pastures, with worm fencing accounting for almost all enclosures. In the Middle Atlantic states and much of Virginia and Maryland, the proportion of enclosed fields was much higher; and while the worm fence remained a conspicuous element on the landscape, the post-and-rail and board fences made steady progress as forest acreage decreased. Throughout most of Pennsylvania and New

York the never too abundant stone fences were no longer being built or
kept in repair; and in some instances they were used as sources for road
paving material.

A detailed examination of the travel literature for New England and
the Maritime Provinces suggests that the stone fence dominated the
scene. But one curious device noted by the Duc de La Rochefoucald-
Liancourt in 1799 along the Rhode Island-Massachusetts border merits
some notice since it was to be repeated much farther west in other wood-
less areas:

> The land is there, for several miles, so sandy and boggy,
> that no stones can be found for fences. On the other
> hand, wood is so scarce, and so costly, that it can be as
> little applied to this use as stones. Yet the fields are en-
> closed with fences, which, to two-thirds of the height,
> are formed with turfs, with cross-bars of timbers above.
> In other places where stones are not so scarce, the
> fences are formed one-half of stones, one-half of wood.

The Puzzle of the Stone Fence

Although it might be imagined that the stone fence was a simple and obvious "response" of man to natural environment in New England, a recent study of fences in a tract along the eastern shore of Union River
Bay, Maine, where there has been little change during the past century indicates that other factors may have been present. Ernest S. Dodge and
Walter Muir Whitehill write:

> There is more clay in the soil of eastern Maine and few-
> er boulders until one gets back a short distance from
> the shore. When the fields were cleared the rocks were
> sometimes built into walls but more often they were
> thrown into large rock piles. And so, in general, in this
> remote corner of New England stone walls did not form
> pasture fences—at least not so frequently as elsewhere.

Great variability in the incidence of stone fences within New England
and adjoining states is shown in the single statistical source on American

fence types available to us—*The Report of the Commissioner of Agriculture for the Year 1871*. This document, based on data collected from 846 counties throughout the nation, was published at a time when the farmers of America were confronting a genuine crisis in their efforts to build and maintain fences around their fields without sliding into bankruptcy. This problem, brought on by the steadily rising costs of lumber, labor and land, was resolved in the nick of time by the development of barbed wire and various types of cheap, mass-produced woven wire fencing. Since the advent of this technological revolution in the 1870s, there has been a sharp drop, and usually a complete halt, in the building of pioneer fence types, so that the 1871 statistics are doubly valuable.

A tabulation of the 1871 data for New England indicates that the frequency of stone fences ran from an average of thirty-two percent in Vermont and a surprisingly low thirty-three percent in Connecticut upward to sixty-seven percent in Maine and seventy-nine percent in Rhode Island, with considerable variation among counties. It appears that nearly half the stone fences of Massachusetts, where their overall incidence was sixty percent, consisted of a combination of stone and wood, a type now nearly extinct. Approximately eighteen percent of the fences in New York, a state settled in good part by New Englanders, were tabulated under the stone category. Although only four percent of the Pennsylvania fences were of stone, the figures were considerably higher in the extreme northeastern counties—again an area that had a large influx of New England settlers. A "fair proportion" of stone fences was reported for Rappahannock, Scott, Albemarle and Fauquier Counties, Virginia, but not one is mentioned south of Virginia, nor west of Pennsylvania, for that matter.

In addition to stone fencing, the board fence was built in great numbers in certain New England counties and accounted for no less than thirty-five percent of the New Hampshire total and twenty-seven percent of Vermont's. In the western, and relatively late-settled counties of Vermont, worm fences outnumbered all others; but in the remainder of New England they amounted to considerably less than ten percent of all fences.

In the Middle Atlantic states both the worm and post-and-rail fences were in widespread use in 1871, but only a small number of board fences are to be found. Farther south (and west) the worm fence comprised almost the totality of the fence population. The board fence ran a quite inconspicuous second in the South and Middle West: but at a time when farmers were desperately casting about for cheap alternatives to wooden fencing, the osage orange and other hedges were well represented in some localities.

A Few Generalizations

From personal observation I offer a few general remarks on the present-day status and distribution of the older fence types of the Atlantic Seaboard. First of all, it should be realized that since the general introduction of machine-made fencing the construction of these earlier forms from local materials has become a lost art—except in certain isolated or retarded tracts and those privileged areas that are the habitat of the gentleman farmer.

Although its numbers and range are sadly shrunken, the worm fence is still the most widespread of the pioneer fences. In New England it is abundant only in the remoter sections of Maine. The relative inaccessibility of large tracts in New Brunswick and Nova Scotia has insured the survival of a goodly number of specimens in those provinces. Confirmation of the former importance of rail fencing in New England is offered in a rather unexpected source—the *Linguistic Atlas of New England,* a fascinating repository, incidentally, for a huge variety of regional fact and lore. No fewer than 25 synonyms for rail fences were recorded in the field.

The Worm Fence Survives

Scattered examples of worm fencing appear throughout the Middle Atlantic states, most often in the mountainous or heavily wooded sections; and, again, in the South Atlantic states the Appalachians and coastal forests constitute an important refuge for this dwindling relic. A large number also remain in evidence in the northern Virginia Piedmont. In the course of detailed field work on settlement features in all sections of

Georgia from 1950 to 1952, data were collected on the occurrence of worm fences, the only significant variety of early fencing remaining in the state. The results indicate a high correlation with inaccessibility, for, with few exceptions, worm fences were come upon only in the mountainous counties of the north or in the isolated Piney Woods region of southern and southeastern Georgia.

The post-and-rail fence, as far as I can judge from infrequent sightings, is on the brink of extinction as a genuine farm fence in the United States, except possibly in southeastern Pennsylvania. It has been revived, mainly as an ornamental device, in some suburban neighborhoods and appears around the pastures of some rural showplaces. The one area where it still thrives with all its original vigor is the lower portion of the St. Lawrence Valley. On many hundreds of the older French-Canadian farms this is the only type of fence in use, and it remains unaltered from the days of Peter Kalm, except for the substitution of strands of wire for strips of wood as the means for suspending the long, narrow split rails between the pair of vertical posts. There are, incidentally, many archaic house and barn types also inviting inspection within this same region.

The current distribution of the board fence, once a popular type throughout much of the Northeast, presents us with an interesting unsolved question. Except for a few estates in the exurban counties of Pennsylvania, New York and Connecticut, the board fence has not survived in any rural area north of the Mason and Dixon line. On the other hand, it remains strong throughout much of Maryland and is quite conspicuous, both painted and unpainted, in the northern Virginia Piedmont and portions of the Virginia Blue Ridge country, fading out gradually toward the southwest. This fence type is, of course, the symbol, *par excellence,* of the landed gentry and the breeding of fine horses and cattle, so that it is hardly surprising to note an especially high incidence in Fairfax and Fauquier Counties, Virginia. What makes the distributional pattern slightly perplexing, however, is the persistence of this fence on many rather humble farms within Virginia, in spite of its virtual disappearance on Northern farms.

The hedgerow as a deliberate method of enclosure remains a rarity on the American scene. Most early commentators attributed this fact to the inability of the hawthorne and other Old World thorny shrubs to thrive under American soil and climatic conditions, but what may have been more decisive was the high cost of labor for setting out and trimming the hedge and the long lapse of time between planting and maturity. There are many places along the Atlantic Seaboard where more or less ineffectual hedgerows have been allowed to sprout along fence lines. On the other hand, the planted hedgerow is common only in the northern Virginia Piedmont—an area, incidentally, which seems to harbor more than its quota of early fence types—but appears sporadically in other sections of the Southern Appalachians.

At the present time the stone fence is commonly encountered only within the confines of New England—and in one tract in northern Virginia. Indeed, no other old fence type can be found with any regularity in New England. It is clear that the stone fence has almost completely vanished in those sections of New York and Pennsylvania where it once flourished, the few exceptions being hardy veterans from another century. Its persistence in southern Loudoun and northern Fauquier Counties, Virginia, is difficult to explain except, possibly, as one aspect of the deliberate pursuit of the archaic and picturesque in a region where such qualities are highly esteemed. Within the more morainic portions of the upper Middle West occasional stone fences in a state of hopeless disrepair are still to be observed, but their crude appearance suggests that little care or skill went into their construction. Even where this fence type persists there are few, if any, new ones being built, and many of the older fences are gradually disintegrating. Indeed, one of the characteristic features of twentieth-century New England is the crumbling stone wall doggedly protecting one dense stand of second-growth forest from another.

Although the temptation is strong to accept the former stone fence region as coterminous with those areas that were densely speckled with glacial boulders or where there were extensive outcrops of bedrock, there are wide chinks in any such argument. The stone fence simply failed to materialize in many portions of the Middle and Southern Appalachians or French Canada where physical conditions were propitious. Such scanty evidence as we have on the early succession of fence types indicates that the stone fence, as we now know it, was not the first response to the need for fencing in the rocky portions of the Northeast. When it was imperative to clear the stones away from a field so as to get at the soil, the rocks could be tossed into large heaps somewhere within the field or into ridges along its edges. The carefully constructed stone fence appears to have succeeded a period of experimentation in brush, stump and rail fences; and once the habit of building them was acquired, it may have become a regional cultural trait, but evidently not one that was clung to too tenaciously.

In passing, it is well to note that the whole question of the place of stone in the material culture of North America merits some speculation and careful work. Why the contradiction in New England between the ubiquitous stone fence and the non-use of stone as a house-building material? Why was stone an important house material in some parts of southeastern Pennsylvania where it was distinctly subsidiary as a fencing material? And why was stone employed so widely for the underpinnings of houses and for chimneys in the Southern Piedmont and Appalachians and almost never for any other purpose?

Do these scraps and patches of information amount to no more than grist for the antiquarian? If we rouse our imagination, they speak to us, though indistinctly now, of those more than two hundred years of struggle to wrest order and security out of a strange continent; and in the where, when, how and by whom of these vanishing fences we can perhaps learn a little more, not only of our predecessors but, indirectly, about ourselves as well. Just as the pursuit of prehistory adds greater meaning to the rhythm of our era, and anthropological dealings with primitive peoples open unexpected windows upon our own minds, so too these early experiments in the technology of space may help us appraise more knowingly the deeper forces underlying contemporary America's varied and quickly shifting rural landscapes. Surely they call for meticulous and far-ranging study before oblivion closes in.

SPRING: SILENT OR RAUCOUS?

J. B. Jackson

Alarmed by the image of a silent spring, a spring without birds, we are reluctant to admit that there can be such a thing as too many birds, birds in such numbers that they can rightly be called a pest. Yet such is the case, and pests of this sort show no sign of diminishing.

The red-winged blackbird and the starling reach a population estimated at 15 million in the southeastern states, and, according to James Fisher, writing in the London *Times,* the greatest nuisance of all is the red-billed guelea, "an African weaver whose nesting colonies may reach 20 million birds, and which is seriously damaging the new open-ranged small crops (wheat, rice, sorghum) of hungry emergent countries all over Africa."

For what encourages these birds to multiply at such a rate is our modern method of "clean" farming. Each year sees more of the rural landscape in every part of the world deprived of its diversity of terrain and vegetation, and vast expanses given over to the raising of a single crop— usually a crop which birds can enjoy eating. Commercial farming thus not only produces an abundance of food for humans but an abundance of birds—and eventually an abundance of pesticides, the more lethal the better.

An alternative to the pesticides (and the dynamiting and the flame throwing and the poison gas) that many conservationists recommend, is a return, insofar as a return is possible, to the traditional diversified landscape where many crops were raised instead of a single one, and where

[from *Landscape,* vol. 13, no. 3, Spring 1964]

there was more variety in the terrain and in the flora and fauna. Where such traditional landscapes persist it is quite true that the bird pest problem can be kept under control.

But it is hard to believe that many commercial enterprises will willingly forego the benefits of large-scale mechanized operations merely in order to control a horde of birds which, in any case, they are sure they can destroy by poisoning. A far more realistic solution is the one tried out in the polder landscape of Holland and in Israel: the deliberate creation in every large commercial farm landscape of artificial wilderness areas, something Artur Glikson proposed many years ago. The artificial wilderness is not only an efficient device for water and soil conservation, for providing recreation areas; it also serves to introduce a biological variety in the farm landscape. Too many birds: the remedy is either to produce the kind of landscape which encourages competition and keeps down the numbers, or else to produce pesticides powerful enough to kill off these hordes, and along with them most of the animal life in the countryside.

A NEW KIND OF SPACE

J. B. Jackson

One useful way of defining cultural geography is to say that it is the study of the organization of space, the study of the random patterns we impose on the earth's surface as we live and work and move about. According to this definition a landscape can be seen as a living map, a composition of lines and spaces not entirely unlike the composition which the architect or planner produces, though on a much vaster scale.

But these spaces are not merely organized in the sense of being arranged into a coherent whole, they are forever being *re*-organized. They encroach upon one another, absorb one another, expand, contract, multiply, disappear. And though this reorganizing of the landscape is always underway, there have been times in history when it has proceeded at an unusually fast pace. One period here in America was recent: less than a century ago. Like most of its predecessors it was entirely without overall direction; it was the product of innumerable private decisions and inspired by a variety of motives—economic, esthetic, technological, political. Farmers, land speculators, politicians, artists, social reformers, all had a part in it. This reorganization took place within two or three decades in the latter half of the nineteenth century, and it is only now, as we in our turn undertake a spatial reorganization of our own, that we are in a position to appreciate its scope. For it is clear that between the end of the Civil War and roughly the year 1887—a span of time as long as that between the end of World War II and the present [1969] —a series of

[from *Landscape,* vol. 18, no. 1, Autumn 1969]

modifications, many insignificant in themselves, drastically transformed the face of America.

An easy and effective way to reorganize space is to put up a wall or a fence. We can thus cut a space in two, give it a new shape, a new character, make it smaller and more useful. But there are two distinct reasons for putting up a wall or fence or any other effective kind of boundary: either to keep intruders out, or to keep your own possessions in. And these two reasons represent mutually antagonistic ways of organizing space—one of them is likely to replace the other.

To give an illustration: in a certain rural landscape there are three landowners, each having about a square mile apiece, who raise livestock. By common agreement they allow their animals to run together; no walls or fences separate them, for the owners have some way of telling their own—either by the breed or by some mark or brand. Perhaps they have arranged to divide the animal crop at roundup time into three proportionate parts. In any event the practice is a satisfactory one; it relieves the owners from having to organize their space.

But along comes a farmer who buys a few-score acres nearby. He immediately plants corn or wheat, and no sooner does the crop start to come up than all the neighboring livestock descend on it and start devouring it. Naturally enough, the farmer demands that the animals be kept out of his fields. The stockmen, however, make what is to them the logical answer, "If you don't want our animals in your fields, put up a fence to keep them out." To which the farmer retorts, "They're your animals. Put up a fence to keep them in!"

Eventually they go to court. The decision of the judge is that it is up to the farmer to build the fence. One reason for his decision is that he is inclined to side with the established setup; but he has another reason as well: if the stockmen were to be compelled to build a fence it would be at least eight miles long, whereas if the farmer has to build one it will be less than a mile in length.

This is a much oversimplified example of the conflict that almost al-

ways grows out of a confrontation between the two kinds of spatial organization, a conflict repeated over and over whenever farmers and stockmen live in the same landscape. Where stockmen predominate, as they
did in the Colonial South and in the pioneer Midwest and West, the law
has it that the landowner must build a fence around his property if he
wants to exclude trespassers. On the other hand, when farmers predominate, as they did in New England and as in time they usually do everywhere, the law says that fences must be built to contain the livestock.

The implications here are wider than may at first appear. Whenever
space is organized for defense or exclusion, its units are likely to be isolated, and much is usually made of the means of access. Door, gate, port
of entry is then important and elaborate. And change, coming as it does
from outside pressure, is something to be feared and resisted. But when
space is organized for containment, a number of specialized areas are
likely to exist side by side; the means of access is easy, and what produces change is the expansion of the contents. Spaces for containment
are therefore apt to grow larger in the course of time, while spaces for
exclusion are apt to remain the same or to become fragmented. The Medieval walled city surrounded by its satellite villages is an example of one
kind of spatial organization, and the modern city, expanding all over the
place, accessible from every direction, is an example of the other kind.

Friction between the two points of view in America was evident as
early as the 1840s, when pioneer farmers disagreed with pioneer stockmen as to whether animals should be fenced in or not; but the issue became acute after the Civil War. One section of the country after another
began to agitate for some sort of control of wandering livestock, for
some modernization of the old customs. The South in particular was un-
happy with the traditional organization of space based on exclusion.
Many of the large plantations were being broken up, and the number of
small farms was increasing. But would-be farmers found to their dismay
that they still had to fence their land to keep out stray cows, hogs and
horses, and for the farmer beginning his career, the building and maintaining of fences was a costly and time-consuming undertaking. Midwestern farmers, moving into the prairie landscape where timber for fences
was hard to get, were loud in their demands for herd or fence laws, and

when stockmen began to improve the breed of their cattle, as they did in
the late '60s, they too demanded the confinement of the less desirable
stock still running free. County after county, and state after state passed
laws that all livestock had to be kept behind fences. Until the invention
of barbed wire in the middle '70s, the whole question of fences and herd
laws occupied an astonishing amount of space in newspapers and farm
journals, and generated strong feelings. "Depending somewhat, of course,
on the locality," Bogue comments in *From Prairie to Cornbelt,* "it
ranged livestock men against grain farmers, town and village dwellers
against farmers, old settlers against new settlers, the well-to-do farmer
against his poorer neighbor, and resident farmers of the county or township against the owners of 'foreign herds'." And what it also produced
was a quantity of excellent writing, chiefly in farm journals, on what we
would now call the organization of space.

The first of the statewide fence laws seems to have been passed in
New York in the early '60s—New England had always had them, of
course—and by 1886 almost all the counties in the Midwestern states and
many counties in the Western states had legislated against free wandering
livestock. But even today it is possible to find counties in the South and
in the Ozarks, as well as in the Southwest, where cattle are still legally
entitled to stray at large and where farmers, in consequence, must fence
their fields.

The general acceptance of these herd laws coincided with and may
have been influenced in some places by, the widespread increase in the
size of American fields. Mechanized farm equipment called for larger,
more regular areas for its operation, and even farmers without mechanized equipment tended to expand the amount of land under cultivation
for the raising of one or two commercial crops instead of the traditional
variety of crops for home consumption. The newer postwar cotton and
sugar plantations in the Mississippi Valley and Texas were likewise characterized by larger fields, and the farms established under the Homestead
Act of 1862 had—most of them—a simpler organization of space. Travelers in the new wheat-raising sections of Minnesota and Iowa after the
war were amazed at the size of the fields they saw. As Horace Greeley
observed, "When people get out into the prairies a feeling of expansion

takes hold of them." He counseled against small farms, and in 1870 he wrote that "the good farmer will have fewer and better fences than his thriftless neighbor."

A new scale was beginning to appear in the American landscape, not only in the open country, but in the towns and suburbs as well. A generation earlier, Emerson had quoted with approval the old New England saying that good fences made good neighbors, but in 1870 a writer on suburban landscaping by the name of Frank J. Scott devoted a chapter in his book to a denunciation of walls and hedges between neighboring gardens. He called them unchristian and unneighborly, and while admitting that outside fences might occasionally be necessary, declared, "That kind of fence is best which is least seen, and best seen through." And quite in keeping with the newer philosophy of containment instead of exclusion, he further cautioned against making the gate or entranceway too conspicuous or formal. In the 1850s Andrew Jackson Downing had decried the practice of putting iron fences and hedges around individual graves in cemeteries, but his criticisms were little heeded in his day; it was only in the '70s that the so-called "lawn cemetery," where no coping or any sort of enclosure was permitted, became popular.

Thus the legal and economic objections to the fence were reenforced by objections of an esthetic nature; taste as well as efficiency demanded changes in the American landscape. Nowhere was this more apparent than in the reformed appearance of the average country road and village street.

Previous to the passage of local herd laws, the livestock had not only wandered throughout the open countryside, they had invaded the villages; cows, goats, horses and pigs delighted in the overgrown margins of the roads and in the greenery around village dwellings and in the village green or park. All of these had to be protected by fences, and the proceedings of many American village meetings of a century ago contain heated discussions about what to do with impounded stray animals, and what to do with their rural owners. Furthermore, the filthy condition of the unpaved country roads encouraged many householders to use them as convenient dump heaps. Here, for instance, is the bitter complaint of a farmer writing to an agricultural journal in the pre-herd-law era: "It is an old-fashioned notion that highways were made to travel in, and belong to the traveling public. A few have discovered the error of this opinion and now employ them for barnyards . . . others use them . . . for deposits of rubbish . . . Others again make cattle yards and pastures of them, the cows helping themselves to their neighbors' cabbages, the horses jumping into their neighbors' wheat fields, the hogs rooting up the grass walk— and all of them terrifying the little school girls on their way to the district school. We are old-fashioned people, however, and prefer clean neat roads, and quiet animals in good sheltered barns or stables!"

And these were precisely the things which the herd laws produced: clean roads and animals safely confined. Towns which Scott had dismissed as "semi-civilized places where hogs are allowed at large" were at last enabled to take down the walls and fences surrounding their gardens and parks. Out in the country farmers began to remove the fences which separated their lawns and yard from the public road, and the margins themselves were trimmed and kept in order.

It was at this juncture that an American institution, most valuable in its time, but now largely forgotten, came to the fore—the rural improvement society. The first of these had been founded in Stockbridge, Massachusetts, in 1853, and its mission had largely been the planting of trees along the streets of the town. But the great day of the village improvement society—there were infinite variations on the name—came after the Civil War. B. G. Northrup, a Connecticut public official, devoted many years to the organizing of such groups, and by 1880 he had more than a hundred to his credit. He also wrote extensively on how to improve and beautify towns; Horace Bushnell appears to have played much the same role in California. But these were essentially grass-roots organizations, initially inspired by a desire to improve the tone of village and rural existence. It was in the 70s that they began to concentrate their energies on beautification, and it was then that state legislatures passed laws to allow the planting of ornamental trees along the right of way. These village improvement societies can also be credited in many cases for doing away with the fences surrounding public parks.

The results of these several activities can still be seen and admired. The sweep of lawn reaching from the house down to the edge of the road or street with its double row of shade trees is so typical of small-town and rural America that we rarely wonder as to its origin, and even more rarely recognize it as a specifically American landscape feature. Yet we are justified in saying that the wide, formally landscaped street or roadway, and the spacious, unfenced lawn both became general only after the Civil War.

Since that time the American road and highway have come more and more under public control; first the alinement, then the surfacing, and finally the margins, were made subject to public supervision and design, and the process of expansion continues. The unfenced lawn is not yet part of the public domain, but it is prescribed in many housing developments and is an essential feature of every public building in the country. Neither of these familiar aspects of the rural and urban landscape could have existed without the passage of the various herd laws or the work of the village improvement societies; both promoted a new kind of space.

It would not be hard to show that the same sort of spatial reorganization took place in American architecture during that same post-war period. Almost every kind of structure and layout in the country, from the barn to the public school to the church to the summer resort cottage, underwent a radical change: interior walls and partitions were either eliminated or perforated by many openings, and the tendency was to increase the size of the structure—to add new units or to expand into the immediate environment and to spill over into the outdoors. The principle of exclusion yielded to that of containment. And once content is thought to be more important than the container, once people are thought to be more important than the institutions to which they belong, then there is little to stop expansion. It is not without significance that during that post-war period a new concept of territorial subdivision developed: of units defined in terms of the natural resources they contained—irrigation districts, national parks, watersheds, forests.

But even the partial evidence drawn from the study of the landscape is enough to suggest the complexity of the spatial reorganization under way, and the variety of reasons for it. Social and technological change produced a new pattern of farm enclosures and a new rural landscape.

Shifts in fashion and taste produced the same sort of change in the design of gardens and cemeteries and parks. Finally, a movement for social reform and improved living conditions, sparked by countless groups of village citizens, produced the same changes on a local scale. Southern farmers, Midwestern homesteaders, landscape architects, village selectmen, all equally inspired, all oblivious of the efforts of others, worked together to reorganize American space.

While it is inviting to speculate about the cause of this remarkable synchronism, all that we can safely say is that there seems to have come a time in this country when all boundaries, all traditional divisions of space, whatever their original purpose, were seen not as desirable forms of protection but as obstacles—obstacles to beauty, to efficiency, to sociability, to change. And when this revelation became general, there was a resolve to do away with those obstacles. So down came fences, walls, hedges, gates and barriers all over the landscape.

What subsequent generations, including our own, have done with this free and spacious environment is another story, and a very different one.

COMMUNITIES

A growing segment of our citizens will not be economically productive except as consumers. In these trends there may very well be possibilities for a resurgence of life in the economically by-passed small towns of America. —Albert Solnit

Every boom of this sort, every sudden and intense exploitation of the economic resources of a region is a kind of exaggerated, stepped-up version of the society of its time. Exaggerated because in new surroundings and new conditions traditional restraints cease largely to operate; stepped-up because a whole cycle from birth through growth, dissolution, and death is enacted in a decade or two. —J. B. Jackson

A large part of our concern for farm and small-town America is just nostalgia: a nostalgia felt not only by small-town expatriates but by life-long citydwellers as well, a nostalgia for a too-placid world which never was. —Robert B. Riley

The villages and small towns of America are not dwindling and disappearing because their values are no longer meaningful, but because they no longer work economically, no longer provide the level of service and amenities that most of us demand. No dramatic violence is being done to rural America. It is withering away because it has little function in modern life. —Robert B. Riley

WHAT'S THE USE OF SMALL TOWNS?

Albert Solnit

After neglecting it for years, we have begun to pay attention to our urban environment. Whether that environment can be truly readjusted to serve a meaningful purpose is open to question, but at least the direction of future urban growth and its objectives are being subjected to a thoughtful analysis. But what about a similar reappraisal of the rural and small town environment in America?

We have always been responsive in this country to the challenges of growth and new settlement. From William Penn to William Zeckendorf, the gospel of "more growth equals progress" has been the popular basis for development in the United States. But in recent decades a great many rural and non-metropolitan communities have had to face the problems of static or declining populations and economies. Such communities are particularly numerous in the Great Plains, in the Spanish-American Southwest, and in the Southern Appalachians. Their economic plight has been studied in detail: the impact of mechanization, of the consolidation of production units, of fluctuations in the commodity market, of technological displacement have been discussed in an extensive literature. One result of this concentration on economic and sociological factors, on the problems of the "culturally deprived," "the disadvantaged," "the retrainable human resources," etc., has been the almost complete neglect of a no less important topic: the relationship of the individual in these areas with his environment.

[from *Landscape,* vol. 16, no. 1, Autumn 1966]

Conventional wisdom, obsessed with growth, has usually prescribed some kind of technical assistance for these communities, which in turn has meant the creation of a climate for investment in industry and commercial tourism, usually by means of public spending on such items as roads and sewers and public buildings, and by means of a campaign to produce a favorable "image." For the truth is, many of these communities have lacked the potential to compete in any serious manner for the tourist dollar or the industrial payroll.

As recently as 1963, the Area Redevelopment Act was largely based on the proposition that each town and county of a depressed area could be saved through local boosterism and boot-strap initiative. The Federal retreat from the notion that every small town could be revived if it was given enough aid possibly dates from the $2 million debacle at Wink, Texas. Elaborate schemes failed to attract developers to renew the center of this moribund oil town, but they enabled a number of the leading citizens, who had sold property to the renewal project, to leave Wink for good. The regional directors of the Housing and Home Finance Agency were likewise relocated. In 1966 a new generation of Federal programmers has found it safe to admit publicly that small towns and rural counties have a limited economic potential. This new attitude was summed up by John L. Sweeney, Federal co-chairman of the Appalachian Regional Commission. He gave the following three imperatives for his program:

"1. Ignore the pockets of poverty and unemployment scattered in inaccessible hollows all over the area.

"2. Concentrate all the spending for economic development in places where the growth potential is greatest (viz: the region's cities).

"3. Build a network of roads so that the poor and unemployed can get out of their inaccessible hollows and commute to new jobs in or near the cities."

Thus, in contrast to the Thoreauvian maxim that man's home should be among the living things of the natural order, the Federal area developers conclude that man should make his home in the *economic order* among the things that offer a living. Both dogmas offer only partial answers to the problem of how men should live. In a have-not or underdeveloped nation, economic development must come first because general conditions of scarcity and lack of productive capacity must be overcome in order to stave off starvation, disease and homelessness. The United States, however, has an economy of abundance and the question of whether economic development is an end or a means was answered by John Kenneth Galbraith:

> Economic development enables us to pay the price; it is why we have development. We do not have development in order to make our surroundings more hideous, our culture more meretricious, or our lives less complete.

So we have a choice. Can we now pay the price of making our surroundings more beautiful, our lives more complete, or must every region contribute equally to the national income? If the goals of the "Great Society" are more than slogans, it would seem that the nation has already chosen the first alternative. What will finally determine our policy is the changing nature of the labor market in America. In the past it could absorb most of the unskilled. But with the rapid rise of technological innovation, automation, and economies of scale, we've reached the point where the government is predicting that 22 to 25 million new jobs will be needed in the next ten years to replace those eliminated by these innovations in the mass production industries alone. These are the industries that have traditionally absorbed most of the surplus population of the small towns and rural areas. These areas in turn have suffered the most drastic reductions in local labor demand because their agricultural and mining processes have undergone even more consolidation and mechanization, with the consequences that their labor input per unit of output has shrunk fastest of all. So the present situation seems to be one where productivity and output are growing faster than employment throughout much of the economy. Furthermore it appears that the nation has gone way beyond any real necessity to keep on increasing the rate of growth of productive capacity and per capita output in many of its basic industries.

Therefore the problems of poverty have become more a problem of distribution than of scarcity. More important, the belief that we *must* have (or *can* have) full employment of the nation's productive age group is beginning to lose its hallowed status. Up to now, a real facing up to the situation has been avoided by shortening the work week and year as well as reducing the labor force. We have increased college enrollment, reduced the high school drop-out rate and made retirement easier, earlier, and more secure. Yet a real confrontation with the question of a non-work society now seems to be in the offing. Evidence of this can be found in the recommendations of the President's Commission on Technology, Automation and Economic Progress, which unanimously urged: That every American family be guaranteed a minimum annual income above the poverty line; that the Federal government should become the employer of last resort for the hard-core jobless, paying them to work in useful community enterprises; and that every young American be offered free education for two years after graduation from high school.

In the same vein, C. P. Snow, the British Minister of Technology, predicted that the advance of science will favor "nonwork" with the result that 10% of the people in the advanced societies will work harder, while the other 90% will work less or not at all. He foresaw a "Brave New World" condition where individuality will become dangerously unimportant.

Since the preservation, not to speak of the development, of the individual personality is threatened, should we not give thought to implementing those suggestions that man abandon machine society and return to nature? Lewis Mumford, in his essay "The Social Function of Open Spaces" (*Landscape* Winter 1960-61) gave the proposal its most persuasive form: "To create a counter movement to the irrationalities and threatened exterminations of our day, we must draw close once more to the healing order of nature, modified by human design."

The controlling phrase here is "by human design," or at least it ought to be. The romantic notion that man is ennobled by being left alone with nature is not borne out by the evidence. Two groups long isolated in their natural environments are the Spanish-American villagers of Northern New Mexico and the mountaineers of Eastern Kentucky. Today both are conspicuous victims of a poverty which for generations has resisted the standard panaceas of social welfare and economic development. Neither group has developed any strong tradition of responsibility for its natural environment; few of their communities have evolved any sense of cooperative effort for long range improvement. The fact that both groups have been engaged in fishing, hunting and farming, so that their contact with nature has been on an immediate individual plane, has not sufficed to produce a society held together by ecological ties. The nature of the society is more existential. They have treated their natural setting as a set of utilitarian objects, to be exploited for subsistence without regard for long range productivity. Nothing has been further from their intention than the romantic notion of preserving nature in her unspoiled state. Though the folklore aspects of these depressed communities have a certain charm, and though a survey by a domestic Oscar Lewis might well reveal a rich inner life among their members, it is nonetheless evident that long isolation with the natural environment has failed to produce an adaptable population. As for such groups as the Mennonites, the Amish and the Pueblo Indians, who have strong communal structure *and* a high degree of harmony with their environment, we can reasonably assume that their admirable adjustments to their surroundings stem from the strength of their social systems rather than from the influence of the inanimate objects around them. We thus come back to the conclusion that it is "human design" rather than the mystic "healing order" of nature that makes life in the country complete or empty.

We can perhaps draw some conclusions about the future of the small non-urban community. First of all, much of the population in many American small towns would be either marginal or unemployable in any city. In their home towns, such people usually have the advantages of roots, family, some kind of standing within a group and often have their own homes. It is clear that they are probably better off at home than in a housing project or furnished room in a strange city, assuming a fairly uniform national level of welfare.

Blackey, Kentucky—a somnolent former coal camp—is an example of such small communities. A resident described it in the Whitesburg *Mountain Eagle* as follows:

> We have at Blackey a compact community of 403 persons—who virtually have been bypassed by the stream of progress and history.
>
> In our population, two groups predominate—the very old and the very young. We do not have a problem of unemployment—it is rather a problem of the unemployable. Less than ten per cent of our population is normally employable. More than ninety per cent of our population is on the draw—from Social Security, pensions, or welfare in one form or another. Thus our chief resource is the human resource.

Someday the very young segment of Blackey's human resource is undoubtedly going to leave. For them there is a pressing need to improve the quality of their education, so that their job market can be the entire nation. Since in our mobile society the shortcomings of schools like Kentucky's are very likely to be visited on the cities that receive their product, proper education can no longer be allowed to be an accident of geography. It will not be enough, therefore, to guarantee a minimum income level; a minimum level of educational quality for all the Blackeys in the United States will have to be guaranteed as well. To do this, there has been a strong trend toward school consolidation in the rural areas of America.

This brings up another emerging fact about non-urban settlements: that the concept of a community as a closely circumscribed, self-sufficient area is no longer valid. Extensive networks of roads and communications have allowed a dispersed population to range over an ever widening radius from their homes. Travel time rather than density of population is coming to be the yardstick for the location of many facilities and services. As a consequence, many trading centers and political jurisdictions established in a horseback era are obvious anachronisms. This has been recognized at long last by the [Johnson] Administration's $5 million proposal to stimulate the formation of rural development districts, comprising (in the words of *Time* magazine) "towns, villages, and sparsely populated counties whose common needs are frequently frustrated by political boundaries."

This measure may help the installation of essential public works and services, but something vital will probably still be lacking. Though programs for raising the material living standards certainly have merit, they do not in themselves bring grace and vitality and meaning into the lives of the people who inhabit the static settlements of the nation. The provision of a sanitary piped water supply is part of upgrading the living standards of any place, but turning on a water tap does not provide one of the recurring satisfactions of living.

Perhaps what is needed to introduce these qualities is an *ecological* rather than a developmental approach to future aid programs. In the sense I use the term, ecological refers to the *designed* interaction between men and their environment. In this definition, men are considered as individuals rather than human resources, and the environment is not composed of natural resources and recreation facilities so much as it is the surroundings in which these individuals live. Furthermore for this eco-system to really be an economy of organisms, there must be mutuality between the individuals within it. If a town or village does not have desirable social qualities, then all the minimum standard improvements and open spaces in the world will not make it habitable—a lesson which recent public housing projects in the cities have made very clear. These qualities are not a matter of following a formula—there *is* no formula. Perhaps in the hope of finding one, the Office of Economic Opportunity (O.E.O.) made Blackey the location of a $200,000 pilot project in human resource development. A community center will be operated in the renovated former coal company store, at an annual cost of $61,000. For that kind of money, of course, a nice new building could have been provided by O.E.O., but that would have made no demands on the abilities and energies of the people of Blackey. Even when jobs are scarce individual responsibility and involvement in a useful activity are necessary for personal growth, meaningful group participation and the creation of a local identity.

But buildings are only a small part of the rural environment. Plants, earth, water and most of the other components of the landscape in Blackey's region have been maimed and mutilated. This landscape must be restored before Eastern Kentucky can again be considered a good place to live in. This also means that the countryside must not only be "beautified" but cared for from now on. In a non-work society the existing site of a mining or agricultural community that was determined by utilitarian considerations could give way to a town design that would reflect the new life style of such a settlement. This is another reason why the natural setting must not only be visually admirable from the road, but good enough to inhabit. The creation of an environment with amenities for living may offer more hope for the ultimate survival of hundreds of small towns than low taxes and cheap labor ever did. This is because more and more Americans are becoming less concerned about the getting of goods than about finding a place in which to enjoy them. Hekki von Hertzen, the main force behind the creation of the sylvan new town of Tapiola in Finland summed up this attitude in a recent interview:

> When we finally caught up we asked ourselves: What
> are we to do with our new affluence? We can't eat
> more. There is a limit to the automobiles and gadgets
> we really need. So I started to persuade my country-
> men that we should build a beautiful and suitable en-
> vironment for everyone. Good housing is not enough.
> We have to counteract the strains and tensions of urban
> life.

The fact that this attitude is gaining ground in the United States is evidenced by the number of families leaving their native cities for places that offer space, scenery and a balance between solitude and society. While the warm climate areas of California and Florida have been most popular with the environment-oriented movers, even areas with rigorous winters have attracted large numbers of former city dwellers. A University of Minnesota study of the Upper Midwest (Montana, the Dakotas, Minnesota, Northern Wisconsin and upper Michigan) revealed that former city dwellers comprised 10% of the region's nonfarm rural population in 1960, with the numbers growing rapidly. In four sample areas having obvious scenic amenities, interviews with over 1400 resident heads of households led the researchers to conclude:

> The amenities of open space and scenery are listed as the overwhelmingly important reasons (50–70% of total response) for decisions to live in these areas of dispersed urban settlement. Although people want a generally open environment, the small acreages they control indicates that they are not concerned about the amount of land they actually control.
>
> The extra cost of dispersal compared with the traditional compact (urban) settlement is generally assumed to be mainly in the form of additional costs of "line services" (roads, transportation, utilities) and facilities. These costs are probably substituted for other luxuries by families in high income brackets; lower income families must compensate by lower expenditures on housing or reduction of other costs. These substitutions are patently *feasible* . . . the evidence suggests that urban dispersal is practical and desired by a significant part of the population.

Much of the year-round population of such places is drawn from seasonal visitors and residents who ultimately settle down on a permanent basis. In the Upper Midwest it was found that once settled such residents moved much less than the rest of the nation.

It is quite common for former city dwellers who consider themselves retired or semi-retired to open small businesses "just to keep busy." As Richard Meier commented: "The sojourners may very well bring enough new enterprise and capital with them to provide full time employment." But before this kind of employment can be generated, an area must develop or restore the amenities that attract them in the first place. Care and enhancement of the landscape thus becomes essential. Turning to Mumford again: "Perhaps the first step toward regaining possession of our souls will be to repossess and replan the whole landscape." To do this properly we will need many more people employed in the semi-skilled and professional aspects of landscape care and management such as forestry, soil conservation, recreation and maintenance, fish and game management and a whole host of associated skills. The objective of planning would be to see that the repossessed landscape was not spoiled by its success. The employment of caretakers for the whole environment would not only be a local economic and social boon, but a national benefit in that the creation and preservation of good space is going to become an increasingly important national resource.

At present, aid programs to rural communities have to choose between make-work projects of temporary consequence or making preparations for the industrial or commercial salvation that may never come. Perhaps their resources could be better deployed to achieve the level of amenity and habitability that might set off the visitor-sojourner-settler reaction. In addition to carrying the seeds of economic revival, an urban population transfusion may very well be the opening for cultural and political change that many closed local societies have so badly needed.

In some quarters mild anguish has been expressed that the dominant national culture is invading and obliterating distinctive regional subcultures. This is a notion fostered by romantic amateur anthropologists, who wish to "preserve" a culture like an insect in amber. A viable culture preserves its traditions while changing living patterns. No culture is static. Even Margaret Mead's New Guineans have made some adjustments to our century. Helping members of a subculture make the essential transitions is much more beneficial than trying to keep them fixed at some point in time. Our inept policy towards the American Indian should make this clear.

We seem to be entering an era where a choice of life styles will outweigh "job opportunities" in the choice of location of a significant part of the population. Perhaps this will bring us to the time when the American dream of an always faster pursuit of happiness will yield to a dream of making living abundant. A growing segment of our citizens will not be economically productive except as consumers. In these trends there may very well be possibilities for a resurgence of life in the economically by-passed small towns of America. It is possible that in these communities we shall finally achieve a design for living fully and naturally, where no man has to detach himself from society to live in the midst of the natural order. Instead we will bring the natural environment to live in the midst of society. If we begin now, 1984 could be a good year after all.

THE FOUR CORNERS COUNTRY

J. B. Jackson

A ten-mile road leads across the desert to a druidical monument: a great stone circle with a concrete monolith in the center. This marks the one spot where four states touch: Utah, Colorado, New Mexico and Arizona. A source of pride to nearby chambers of commerce, it is, in fact, one of the least visited (and least worth visiting) sites in the whole area to which it gives its name. Within a hundred miles of the Four Corners lies much of the wildest scenery we have: Monument Valley, the Natural Bridges, the Goosenecks of the San Juan, the beginnings of the canyons of the Colorado—not to mention the prehistoric ruins on Mesa Verde or in Chaco Canyon. Actually the Painted Desert, the Petrified Forest, even the Grand Canyon are not far away, and in a sense are part of the region. For one of the peculiarities of the Four Corners Country is that while it has a center point—that marker inaugurated by a quartet of governors—it has no boundaries. You can define it pretty much as you like: in terms of uranium or scenery or potsherds or rainfall.

Speaking as a tourist who has merely traveled there many times and over many years, I have always thought of it as a landscape where almost everyone else was, like myself, in transit; where no one really put down roots and was forever on the move. I have thought of it as a landscape where a special relationship—tenuous and fleeting but nonetheless real— prevailed between the enormous emptiness and the people who passed through it. There were reasons for my feeling this. The geographical characteristics of the Colorado Plateau (which is by way of being the core area)—its profusion of canyons and mesas and buttes, its altitude and its extremely dry climate—discourage permanent settlement throughout most of the area. The Navaho, who have been here more than six hundred years, have never really stopped moving; neither townsmen nor farmers, they wander from one valley and plain to another, following their sheep and goats. Spanish explorers passed through, but all that they left was a scattering of place-names. Only the Hopis, isolated on their extraordinary mesas to the south, have taken root as farming people do. In the course of centuries they have created their own miniature landscape in the midst of the much larger one; yet even Hopi farms have a disconcerting way of shifting location. Starting about one hundred years ago, Anglo-Americans infiltrated the region. Where there was a flow of water the Mormons built neat brick villages, laid out fields and planted orchards, but everyone else became wanderers of one kind or another: prospectors, explorers, trappers, stockmen. It has always seemed to me typical of the transient spirit of the country that the town of Grants, N.M., now the center of much uranium milling, should have been named after the tent which the Grant brothers put up to accommodate travelers.

This footloose way of life derives for the most part from a lack of moisture. Except in a few river bottoms or where there are seeps at the edge of mesas—as in the Hopi Country—the Four Corners Country is certainly not meant for farming; and with a traditional technology you could not have a permanent population in villages and towns without an underpinning of agriculture. This has always been grazing country, where ranchers and herders have to keep moving to find grass and water for their livestock—up into the summer mountain, down again to the sandy washes after the rains. It has in consequence always been about as sparsely populated an area as we have. It used to be, before the last war, that you could travel fifty miles or more across rolling sage and pinon country, bounded by endless mesas, without seeing any human habitation. Finally you came to a trading post with a cluster of Navaho hogans like enormous mud-pies, saddle horses and wagons in the shade of a cottonwood tree. It takes some fifty acres here to support a cow; sheep and goats fare better. Railroads were (and still are) a hundred miles away; paved roads were scarce, but so were billboards. Pale mountains with the

outline of hats or chimneys or saddlehorns, floating just above the horizon, served as landmarks for a day's travel; the sound of sheep bells (and
few of them) was all you heard.

As for the few small towns, strongholds of Mormon self-sufficiency
and virtue, they lived chiefly for farming, and as a sideline served the out-
lying ranchers. Cavernous General Merchandise Stores—ceilings dangling
with harness, kerosene lanterns, dishpans, fly paper—sold Levis, cakes
of salt, dehorners and volcanic oil for chapped teats. At shipping time in
the fall there was an amateur rodeo. Main Street was a quiet tunnel of
cottonwoods, the weekly newspaper a mass of rural social notes. The existence of a soft mellow rock in some remote canyon, of a trickle of oil
near a river, were matters of interest only to cowboys and herders. But
there was always a slight movement; tourists on their way to some spectacular destination, trucks headed across the continent; a horseman or a
Navaho wagon raising a minute cloud of dust somewhere miles away; and
simply by being in motion yourself you felt that you belonged.

Then in the '40s the economy of the Four Corners Country underwent an abrupt change. Everyone had known that there were gas and oil
here, but few had bothered to exploit them; the market was too far
away. Everyone had also known that the Colorado Plateau contained
uranium ore; the Indians painted themselves with the yellow earth, and
in 1913 several tons were shipped to the Curies in Paris; but who supposed that it had any commercial value? It was in 1941 that the engineers of the Manhattan Project started the demand. A small wartime
boom ensued. A second, more substantial boom was sparked in 1951
when the Atomic Energy Commission initiated a long-range uranium
buying program. Last year [1959] the Agency slowed down its rate of
purchase while extending the program to 1966, and this decision may be
said to have marked the end of the boom—at least for the time being.
During those same post-war years the oil and gas deposits of the San
Juan Basin—the fourth largest in the United States—were first exploited
in earnest. And finally construction was begun on the Glen Canyon Dam
on the Colorado and on the Navaho Dam on the San Juan, and several
important pipelines were laid.

So the boom had actually more than one impetus; not only uranium, but gas and oil and the Corps of Engineers all played a part. In many ways it must have been like every other mining boom. A half explored, little populated region twice the size of New England was suddenly overrun with eager prospectors; there was the usual epidemic of claim jumping, fraudulent promotion, involved lawsuits, swindling, and fortunes were made overnight. There were rushes by jeep, bulldozer and helicopter to newly opened fields; mysterious killings and disappearances. Discouraging rumors of solar energy being about to replace nuclear energy, encouraging rumors of impending wars and an unlimited demand for uranium ore, were topics of debate in every bar and filling station and hotel lobby.

And after the prospectors came the mining companies, the construction companies, the road crews, the pipeline crews, the housing promoters, the men looking for jobs. No one could begin to keep track of the growth in the population. San Juan County in northwestern New Mexico grew by 190% in eight years. In 1940 the Mormon village of Moab, Utah, had 880 inhabitants. It now [1960] has 5,000 and wonders whether it has water enough for a new golf course. In 1940 Grants (lumber and ranching) had 800 inhabitants. It now has 15,000 and calls itself the Uranium Capital of the World. In 1940 Farmington, N.M., had 2,000 inhabitants. It now has 23,000, and until the boom came to an end it was dreaming of a fourteen-story office building, just as Grand Junction, Colorado (a competitor for the title of Uranium Capital of the World) was dreaming of an office building that revolved. Now they find it hard to get even a Federal Housing Administration loan. Page, Arizona, a model town as the the Department of the Interior understands the term— wide, curving streets, hospitals, schools and parks—at the site of the Glen Canyon Dam, has more than 3,000 inhabitants; it is too new to be on most road maps. Even the Navaho reservation has been affected; if its population has not noticeably increased it is enjoying for the first time a taste of collective well-being. Window Rock, the Navaho capital has a modern housing project, sadly like all other housing projects. Roads are being built throughout the reservation, and no hogan is entirely out of reach of the fleet of paddy wagons operated by the Navaho police.

Undoubtedly a boom, and its subsequent deflation, can be fascinating to study, and some day the story of this one will be written. My own interest in it, however, is strictly limited. All that I really care about knowing is this: have these last eight years drastically changed the landscape and the quality of the Four Corners Country, and if so, how? Can anything so indefinable yet so precious as the atmosphere of a place survive the sudden impact of thousands of newcomers, millions of dollars invested in construction of every kind, and a transformed economy?

The answer in the case of the Four Corners Country is briefly: yes, it can.

Every boom of this sort, every sudden and intense exploitation of the economic resources of a region is a kind of exaggerated, stepped-up version of the society of its time. Exaggerated because in new surroundings and new conditions traditional restraints cease largely to operate; stepped-up because a whole cycle from birth through growth, dissolution and death is enacted in a decade or two. The Gold Rushes in California, Colorado, and the Yukon were caricatures of 19th Century America. Initiated and to a large extent kept going by small independent adventurers, they were epics of rampant individualism. Those who struck it rich spent most of their money on the spot, and spent it in conspicuous ways. Single for the most part, they demanded gaudy surroundings for their leisure. Even the most temporary of their camps were gingerbread replicas of eastern cities, complete with expensive mansions, opera houses, daily newspapers and municipal politics. For the 19th Century relished urban life, and urban life was what the scores of Virginia Cities, Bonanzavilles and Silvertons reproduced with a grotesque verisimilitude. No less typical of the 19th Century was pride in the "Conquest of Nature." In the Old West, this conquest assumed a particularly brutal form. The extracting of the ore, its processing, even its transportation, entailed the wholesale defacement of mountainsides, streams and forests. When finally the ore was exhausted so was the landscape; all that was left was a collection of preposterous ghost towns falling to pieces, and a maze of track, mine pits, slag heaps, dammed streams, stumps, and increasingly terrible roads. An indication of how our point of view has changed since then is our present tendency to find these abandoned countrysides picturesque.

Since those days we have had oil booms in Oklahoma and Texas and

California, and I suspect their respective landscapes, insofar as they still exist, would tell us something worth knowing about the early decades of this century. But the mid-20th Century has now produced a boom landscape of its own, and in most every respect it is unlike its predecessors.

Mining for uranium (or for that matter, drilling for gas and oil) is not a poor man's game. You do not pan for uranium as you might pan for gold; you look high and low for where it might be found, and then you may well have to dig it out from under a hundred feet of rock—a costly job in itself; and once you have the ore you must haul it to a mill, perhaps two hundred miles away, where it is processed. A good grade of ore is ¼ of 1% uranium; out of a ton you may recover four pounds of "yellowbrick"—the commercial product—which is worth about $7.40 a pound. Millions of dollars have been invested in the mills, and that is why the chief participants (and the chief beneficiaries) in this boom have been, not rugged individuals, but well-heeled corporations.

This is another way of saying that the boom population of the Four Corners Country, unlike that of the Klondike, was largely composed of wage earners. Anyone who knew the area in pre-war days is properly amazed by its present prosperity. But even during the most expansive years it was prosperity of a very respectable, cautious kind. No "palatial residences" were built. There are air-conditioned trading posts and splendid bowling alleys and supermarkets and new branches of Sears; but no branches of Nieman Marcus, no modern equivalent of Gold Rush Grand Imperial Hotels with ballrooms and imported chandeliers. Around the few nightspots there is no jam of foreign cars. Compared to the sleepy villages they once were, the towns are lively and brilliant; the new highways leading into them are lined by super filling stations bedecked with pennants, the sprinklers wave to and fro on the motel lawns, and the trampolines are full of small bouncing figures. A gay and welcome spectacle after the harsh solemnity of the desert, but even at the height of the boom, except for a surge of traffic when the drive-in movie closed, the streets after dark were dead. The only reminder of the old restless tradition were the few Navahos, lounging with their peculiar, long-legged stance, outside the bus depot. *They* at least keep moving.

Wage earners and family men. The permanent residents, the business and professional people in the towns, profited by the flush times to build themselves ranch-style homes in new subdivisions. But more than three-quarters of the labor force at the height of the boom was employed in construction work—dams, roads, mills, pipelines, houses, etc. These men, the most transient of all workers, had no thought of buying a house. Yet in these new towns there was none for rent.

The same dilemma must have faced the newcomers to Cripple Creek. There the solution was for someone to build rooming houses, small dwellings and hotels—expensive and flimsy, but providing accommodations of a sort and contributing to the citified appearance of the place. But here again the mid-20th Century boom found a procedure of its own: the temporary population lived in trailers.

Much the most distinctive feature of this new landscape is, to my mind, the presence of immense numbers of what the industry elegantly calls "mobile homes." I doubt if anyone knows how many of these there were in the boom days; no chamber of commerce ever deigned to count them; but they must have numbered in the tens of thousands. They were (and to a lesser extent still are) scattered everywhere: in empty lots, in farm and ranch backyards, along highways, out on the range, in canyons; singly, in batches of four or five, grouped in villages of a hundred or more. Most of Page, Arizona, is composed of trailers, neatly assembled in parks; and during a six months' strike among the construction workers on the Glen Canyon Dam a thousand trailers moved away, though most have since come back. Modern trailers, of course, have little in common with the small, two-wheeled contraptions designed for campers. Ten feet wide, forty or more feet long, they can cost as much as $12,000, though the price of the average is probably half that. Since they are usually hauled by trucks at the rate of 35 cents a mile they spend little time on the open road. Most of them contain two bedrooms, a living room, kitchenette and bath, though double-deckers can sleep eight persons. Generally speaking, they are triumphs of light, durable construction, ingenious planning and compact convenience. Trailer literature, especially trailer periodicals—there are at least seven—have much to say about the refinements of trailer life: trailer interior decoration, trailer landscaping, even trailer cooking. There are "period" trailers on the market, and trailers which come with "Danish Modern" furniture.

Thousands of prosperous trailer owners—mostly retired people—are probably interested in these topics, but I think it is safe to say that the great majority of trailer inhabitants are young couples without much money. Certainly few of these niceties were in evidence in the trailer courts of the construction and pipeline crews I saw in the Four Corners Country. To these persons, trailers are essentially dwellings on wheels, ready to move out at a day's notice to some other job. There is a distinction, sometimes overlooked, between the resort-retirement kind of trailer living in carefully designed and regulated parks, and the on-the-job kind of trailer living for a matter of months in the cheapest location available. The industry, of course, wants the public to think of trailers as socially above reproach; but, at the risk of appearing contrary, I must confess that the trailer in its dressy resort guise always strikes me as pretentious and synthetic, whereas the less imposing trailer, bleak and untidy though it often is, represents as I see it a sensible and efficient response to certain conditions of work.

Under any conditions, however, trailer life can be very complex and confusing—for the community into which a trailer moves as well as for the trailer inhabitant himself. He is a home owner who is not a land owner nor a tax payer; a resident who is often excluded from residential zones, a transient who must always live among other transients. For trailers are not self-sufficient; they must stay where they can get water, electricity, sewers; that is why there are trailer courts. So wherever you have an influx of trailers there comes into being a number of small, more or less isolated neighborhoods with their own ways of thinking and behaving; and almost invariably there is friction between town and newcomers.

Workers' trailer courts are usually small and compact—not to say crowded—and no money is wasted on beautification. Few courts in the Four Corners Country contained more than a hundred units, and the average I should say was between twenty-five and fifty. This would mean a population of between one and two hundred. As I mentioned, some of the courts are clustered around or near existing settlements, but in this sparsely populated region the larger number are located all by themselves far out in the open country—near the mine or mill or construction project. These are by far the most interesting to see: remote satellite trailer camps, a half-hour out on a dirt road in the range or desert, set in the midst of a gigantic landscape of rock and sunbleached grass. You drive through the emptiness and all of a sudden there it is: a tight, orderly cluster of fifty or more trailers—pink and brown and turquoise blue and yellow, aluminum roofs gleaming in the sun, television aerials like a silvery web floating above them; and not a sign of transition between them and the surrounding landscape. I remember in particular a camp situated in a wild and beautiful canyon of dark red rock in Monument Valley. No more perfect setting could have been devised for those simple, cleancut forms, geometrically arranged with their brillant colors.

Whenever a new construction project is announced somewhere in the Four Corners Country, either the construction company or a promoter (usually operating on a shoe string) leases land from a rancher out where a camp would be profitable and convenient. The promoter drills for water, bulldozes a main street and a few laterals, digs septic tanks, hitches up to the power lines, and builds a public toilet out of cement blocks. But he is not ready for business until he builds the cement block building which is to house the camp laundry. This is the basic social instituion; it serves as a gathering place for the women, it is where mail is distributed, notices are posted, and where the pay telephone is installed. It is also a center of quarreling and disagreement, the place where management has to step in and show its authority. The laundromat eventually expands to include a recreation room for square dancing and the showing of home movies; but even in its basic form it is essential. That is why the humbler camps are usually called "Jake's Trailer Court and Laundromat" or "Canyon View and Washeteria."

Such is the average community: twenty to fifty trailers parked side by side on lots twenty feet wide (forty dollars a month); a wide dusty main street littered with tricycles and portable gas tanks; the public toilets and laundromat and managers' office somewhere near the center. Wash fluttering, children playing, morning glories growing on strings. But in time improvements are made. A playground is bulldozed out of the stones, overnight trailers have their own section, so do couples without children. Because the population is small and not very prosperous, no permanent business finds it worthwhile opening near them. A dealer in new and used trailers moves in down the road; a Pentecostal church or mission may possibly establish itself nearby in a trailer of its own. When several trailer camps grow up in the same neighborhood—Milan, N.M., was once composed of thirty-five of them and little else—something like a larger, more complex community evolves. But it will be no more permanent than its component parts; the streets are not paved, the church is housed in a government surplus building, the movie is a drive-in movie: altogether a dreary place.

And, of course, when the boom slackened, when the large construction projects were completed, the trailer courts started to shrink and finally to vanish. The owners who were in operation to make a quick profit, salvaged almost everything they had. All that remains are patches of brown grass, once lawns three feet wide. The weeping willows and the Chinese elms, planted next to doorsteps quickly die. A winter's wind eventually wipes out the last traces, and "Golden Acres Trailer Ranch and Laundromat" is one with Nineveh and Tyre.

There is much resentment abroad against trailers and their inhabitants. The ugliness of such camps as I have been describing is the source of most of it. I am not certain that a straightforward array of decently spaced trailers in camps efficiently run is not preferable from many points of view to the over-landscaped, over-planted, over-organized resort trailer courts. But there are other than esthetic objections: real estate operators, zoning officials, property owners all have reasons for wishing to keep trailers out of the community, and citizens in general feel that they are a burden on public facilities which cannot expand to meet a temporary need. Others criticize the rootlessness and civic irresponsibility of trailer existence: the loud behavior of trailer inhabitants on Sunday nights, their tendency to pull out of town when bills are due. I daresay each of these objections has a basis. Nevertheless, I have the feeling that neither the critics nor the defenders of the trailer have tried to recognize the trailer for what it is: a new and undefined kind of dwelling, a symptom of far-reaching changes not only in family life but in the evolution of our towns and cities.

Just the same, four million Americans are now [1960] living this way, and not all of them are construction workers. The thing is, trailers and even trailer courts are very concise expressions of certain important trends in American life—traits which may or may not be desirable but which are there.

Mobility is of course the most obvious of these. As a society we are on the move, and every time we move we leave a dwelling behind a little worse for wear, a little nearer to being part of a slum. Several years ago in this magazine, Edward Price of the University of the City of Los Angeles analyzed the decreasing longevity of American dwellings largely brought about by this increasing mobility. At the extreme of mobility—which there is no reason for thinking we have reached—he foresaw "temporary houses, which after a few years could be junked without need for redecoration. The landscape would conform quickly to the function of the moment." He prophesied the evolution of two distinct types of building: temporary dwellings suited to the needs of the mobile element, immobile buildings to house the elements of social continuity—schools, courthouses, churches, etc. Professor Marshall of Columbia went even further at the First International Seminar on Urban Renewal held at The Hague. Too many people," he declared, "including town planners, try to build buildings for all eternity. We should design structures, perhaps whole cities, to be written off quickly." Actually, the trailer is a repudiation of this idea of Marshall's, just as it is a vindication of Price's forecast; because *the trailer cannot be mobile without there being somewhere a fixed, permanent center,* social, economic and utilitarian.

It seems to me that the trailer expresses still another important American tendency: the flight to the suburbs. I suggested earlier that the 19th Century boom landscape illustrated that century's love for the city. The mid-20th Century landscape—if the Four Corners Country is any criterion—illustrates our own almost diametrically opposed desire for a new kind of community; smaller, less complex, less impersonal, where relationships are more spontaneous and familiar; a community from which the over-organized world of daily money-making is rigidly excluded. The trailer camp epitomizes this anti-urban, anti-formal utopia, with an additional advantage which the conventional suburb can only envy: the paternal figure of the court manager, symbol of ownership, continuity and authority, who settles all misunderstandings, listens to all grievances and organizes surprise birthday parties in the recreation hall.

Lastly, there is reason for thinking that the trailer is a logical step in the evolution of the American one-family dwelling. In the past one hundred and fifty years we have seen it lose one traditional function after another: the economic function (as place of work) the religious function, the educational function, and finally, with the advent of community recreation, the social function. The new home, to quote Lewis Mumford, "is primarily a biological institution and the house is a specialized structure devoted to reproduction, nutrition and nurture." And to quote once more from this magazine, "the modern dwelling is now thought of as a shelter, a shelter so temporary that its appearance and location and permanency are matters of small account to its tenants . . . It is a shell, temporarily occupied, associated neither with the past nor the future." The trailer has merely carried the stripping process one step farther: it has relinquished all attempts to relate actively to its (temporary) environment, or to express economic standing.

It hardly seems necessary to underscore the parallels in the boom landscape of the Four Corners Country: the temporary trailers grouped around the permanent laundromat and recreation room and management; the scarcely less temporary trailer communities grouped around the permanent town and service center; that clearly marked point off in the desert around which revolves a vaguely defined territory. The symbolism which comes to mind is that of nuclear energy—two ellipses revolving around the atom; not too inappropriate for the uranium stronghold of America. Will other new landscapes evolve in this same manner—mobile with permanent centers of energy and purpose? Time will tell; but I thought I saw a sign in Page; there all thirteen churches ("elements of social continuity") were grouped close together on one street in the center of town; an amazing collection. Yet in a shifting, amorphous community how can churches any longer hope to be located in the midst of their parishioners?

Transiency, mobility; they are today a characteristic of the Four Corners Country as they were two decades or two centuries ago. And for all that intensive exploitation of mineral wealth the damage to the landscape is slight; by comparison with the damage done in the Gold Rush mining days it is no damage at all. There is no smoke; there are no rust-stained rivers, no slag heaps, there is no deforestation. Oil fields have long since ceased to be the diabolic messes the public still imagines them to be: bristling with rigs and derricks and towers, spouting fire and smoke, reeking of decay. Pipelines lead straight across country, cutting through hills and bridging rivers and canyons, but they are inconspicuous, and, aside from a narrow scar a few yards wide, soon overgrown, the land remains untouched. Much the same holds true of uranium ore mining; among land formations as tremendous as these a fresh surface of rock is scarcely noticeable. Reclamation dams built in a chaos of dust and explosion and bouncing trucks and slovenly trailer camps soon become the shores of solitary lakes almost as inaccessible as before.

How are we to account for this difference in the treatment of a countryside? Is it only a matter of improved techniques of exploitation, or have we really learned to behave decently and not to remove what we cannot replace? Was the 19th Century barbarism a symptom of a destructiveness which we now express merely in a different manner? Or is perhaps this particular landscape tougher and more resilient than most? I don't know the answer: I do know that in the Four Corners Country the old solitude is returning. Already towns are seeking to attract travelers once more. In roadside cafes the talk is no longer of mining but of truckers' hauls and roads and distances. Now possessed of pickups, the Navaho travel farther and farther afield, deserting the trading post in favor of supermarkets in the towns; but they are still on the move. That is why in retrospect the House Trailer People seem to belong here. They are wanderers in a landscape always inhabited by wanderers. They never settled down. The way they came out of nowhere, stayed awhile and then moved on without putting down roots, without leaving more than a few half-hidden traces behind, makes them forever part of this lonely and beautiful country.

CITIES OF THE PLAIN

J. B. Jackson

That sizable and very flat portion of the United States stretching
roughly between Missouri and the Rocky Mountains, scorned by travelers when it cannot be avoided—and it rarely can—may sometime in
the not-very-distant future become one of the newest and most vital
landscapes in the country.

Not that it appears destined to undergo a population boom; quite
the contrary. The Great Plains region is in fact growing at the slowest
rate in the nation and three of its states have actually a smaller population now [1965] than they had in 1930. One reason for this is the
decline in the number of farms. In 1939 there were 800,000 of them;
now there are 500,000, and economists predict that in another generation this number will be reduced by half. The only farms which can
flourish in the wheat belt are those of a thousand acres or more, and
these of course are highly mechanized.

As a result of this, a great number of once flourishing farm villages
in Kansas, Nebraska, the Dakotas and northwestern Texas and Oklahoma are slowly but surely dying, to become latter day ghost towns, remote from the larger highways and all but unvisited. And even well
established rural centers, towns up to 10,000 inhabitants, are having a
hard time surviving. A century or more ago, when many of them were
founded, all that was needed to keep a town prosperous was a railroad
station, a county courthouse and a hinterland of productive farmers.
Now the railroad line scarcely counts, there are fewer farmers and

[from *Landscape*, vol. 14, no. 2, Winter 1965–66]

those who remain think nothing of driving 50 or 100 miles to do business in the nearest large town; and as for the courthouse, serving as it does a dwindling population, it is an anachronism.

What is the answer? Rural sociologists seem to agree that a town ought to have at least 10,000 inhabitants in modern America to be viable. Only a town of that size or over can give us the services and amenities we want: a hospital, a high school, a newspaper, a radio or TV station, a sufficient number of churches of various denominations, supermarkets, etc. These we must have if the community is to attract industry or hold on to what it possessess.

These are precisely the size towns which are now lacking throughout the Great Plains. There are cities—very few of them—and crossroads settlements with a filling station and a general store. But generally speaking by current American standards the region is an underprivileged one, inhabited by well-to-do and energetic people who have no local daily newspaper, no local high school, no local radio or TV station and no local contact with the outside world.

Here is certainly a challenging field for the environmental designer: the creation—often on the basis of existing towns—of modest but versatile urban centers and the revitalization of the surrounding rural landscape; the design of small cities up to 50,000 furnished with an industry, to serve a well-established, homogeneous population—just as the Midwestern County Seat served its smaller and simpler environment so well in the 19th Century. Planners and prospective planners—and for that matter schools of regional planning—weary as they must be of tackling the problems of multi-million-dollar centers, expressway systems, deteriorating downtown areas, chaotic suburbs and indigestible industrial complexes, can seek renewed inspiration in the fact that a vast and potentially important area of the United States, hitherto ignored by urbanists and landscape architects, is waiting for their proffered solutions.

The usual reaction to the infinite emptiness of the Great Plains is the natural one: who would willingly continue to live here if there were an alternative? But the environmental designer, to be true to his calling, has to ask himself a further question: how can the region be made livable and a place where people will want to stay? And one of the answers— though only one—is the bringing into being of attractive and prosperous centers with some of the less complex features of modern urban existence.

NEW MEXICO VILLAGES IN A
FUTURE LANDSCAPE

Robert B. Riley

Driving west along U.S. Route 66 through western Oklahoma and the
Texas Panhandle, one crosses a land that grows steadily higher and drier.
Some hundred miles into New Mexico the mountains first appear—not
the soft rounded green hills one left behind with the Ozarks, but abrupt
angular shapes, blue on the horizon. These are the Sangre de Cristo and
Sandia and Manzano ranges—southern outriders of the Rockies. Just east
of Albuquerque the highway climbs up into Tijeras Canyon, six thousand
feet above sea level. From here, as the mountains and the motel signs and
the gas stations and the souvenir shops crowd in on both sides, one can
continue westward on 66 and descend into the valley of the Rio Grande.
Or, one can turn off north or south onto New Mexico Highway 10. To
the north, the roadside is dotted with a few old houses, more new houses,
and abundant signs advertising mountain homesite developments. To the
south, however, the road passes for twenty miles through National Forest
land. Signs and houses are few, located on isolated pockets of private
land. The villages, almost deserted, have Spanish names—Cedro, Yrissari,
Escabosa. Farther south, at the end of the Forest Service holdings, the
landscape suddenly opens out. On the right, the mountains, still parallel
to the road, are now some five or more miles west; on the left the mountains disappear altogether, and one looks eastward into the Estancia Basin
—a great plain of rangeland and salt beds. Along the road more villages

[from *Landscape,* vol. 18, no. 1, Autumn 1969]

appear. All have Spanish names too: first Chilili, and then, as one crosses from Bernalillo County into Torrance County, Tajique, Torreon, Manzano and Punta de Agua.

Just south of Punta de Agua the road turns east, leaving the mountains behind, and then later south again and on into the town of Mountainair. From the last Spanish town on to Mountainair the pattern of man's presence on the land changes. From Chilili to Punta, few habitations exist between the villages. In the villages themselves, the buildings are the color and texture of the land: walls of adobe, plastered, or exposed and crumbling and washing back to mud; or of local rough stone, or poles intersticed with earth plaster. The peaked tin roofs are rusted to a deeper shade of that same red earth tone. Beyond Punta the remaining clumps of pinon and juniper vanish—cleared for farming or grazing. Isolated ranch buildings appear. These are Anglo structures, not Spanish. They are of weathered wood siding, sitting near occasional clumps of cottonwood trees and rickety windmills. Many are deserted. They appear not of the land at all; with their worn gray boards and angular forms they stand in sharp contrast to the silvered green of the range—aloof, ghostly, melancholy. Their weathered disrepair and the sad creaking of the windmills bring to mind Auden's "ranches of isolation."

The village of Manzano is in the Upper Sonoran Life Zone (in Merriam's old classification), which covers most of the state. The trees are juniper and pinon, a small conifer which, burned as firewood, gives off a pervasive smell that is one of the most memorable and haunting characteristics of the New Mexico uplands. Where the trees have been cleared and the land cared for, there is range grass. Cactus is scattered through woods and grass alike. At Manzano, instead of following the paved highway south and east into the Estancia Basin, one can turn west on to a Forest Service road that climbs steeply up through the Cibola Forest to the lookout tower atop 9,368-foot Capillo Peak. Along this route, the pinon and juniper give way to Ponderosa pine, in the Transition Zone. In the last few hundred feet, trees of the Canadian Zone appear—Douglas fir and Engelmann spruce. From Capillo Peak itself, one can look down into two different worlds. To the west, and nearly a mile below, is the valley of the Rio Grande, a harsh angular land of mountain, mesa, and arroyo characteristic of much of the Southwest—a sharply delineated landscape of changing color and shadow. Eastward, the high plains roll off to the horizon—the few small mountains appearing almost like islands in a great unchanging sea of range and farm land.

The eastern base of the mountains, along which Torreon, Tajique, Manzano, and Punta lie, is the frontier of the ecologist's "true West." Historically, it has been a cultural frontier as well. When the Spanish first arrived here in 1605 they found at least four Salinas pueblos: Abo, Quarai, Tajique and Gran Quivira. These were far eastern outposts of the sedentary Pueblo culture; the grasslands to the east were the land of no-madic Apache and Comanche. The Spanish established a mission at each pueblo. Neither pueblos nor missions could long withstand the pressure of the plains tribes. By the time of the Great Pueblo Revolt of 1680, which drove the Spanish out of New Mexico, both Spaniard and Indian had already abandoned their settlements east of the mountains and had withdrawn westward into the safety of the Rio Grande Valley. The Spanish returned with the reconquest of 1692, but the Salinas people were ab-sorbed into the Rio Grande Pueblos. Permanent Spanish settlement be-gan in the early nineteenth century. The land grants of Tajique, Torreon and Manzano were awarded in 1834, 1839, and 1841—all during the brief rule of this land by the Mexican republic. The four towns, and a few hamlets now gone, survived the Indian attacks, partly by playing off the Comanche against the Apache. After the Civil War, Anglo penetration westward into New Mexico became intense. It overran a much weaker Hispano eastward flow and almost all the Spanish-speaking settlements east of the mountains withered away. The four towns along the Man-zanos again survived. Although today [1969] their population has dwin-dled to perhaps a fifth of what it once was, the towns are still Hispano to their core. Where English is spoken it is as a second language.

The villages had a non-specialized subsistence economy: cattle and sheep raising, small farming, timber cutting and stone quarrying, a little retail business. The Anglos, Texans largely, brought with them large-scale market-oriented cattle and sheep grazing, an economy that prospered greatly with the railroad's arrival. The region did well enough for several decades. The general depression of the nineteen-thirties was partially offset by the concurrent expansion of dry land bean farming—a minor boom such that by 1940 the Estancia Basin was being proclaimed the "pinto bean capital of the world." A severe drought began in 1941 and lasted sixteen years. It wiped out the bean farming. From the early fifties on the story has been a commonplace one: the sure and steady decline of a rural county that depended upon agriculture and the railroad for its jobs.

The Anglos brought other things with them, new financial and legal institutions for which the Spanish-American, from custom and language, was unprepared. The Anglo judicial system was drastically different from Spanish codal law. The communal lands of the original grants had gradually been converted into family holdings and the system of land inheritance divided the holdings into ever smaller plots unsuitable for large-scale ranching. The history of Anglo-Spanish conflict in the Southwest has been written many times. It is enough to say that here, as usual, the Spaniard was from the start the loser. Other factors hastened the villages' decline. Irrigation water is relatively plentiful in the Estancia Basin. Westward, higher into the foothills, it becomes progressively scarcer, and the soil poorer and thinner. The railroads, so important to the country, served all five sizable Anglo towns on the plains, but bypassed the villages completely. Highway Ten remained unpaved until 1958. The spreading fingers of Albuquerque's phenomenal growth in the forties and fifties never came near the villages, or even in their direction.

Less than five hundred people live in the villages now [1969] —mostly school-age children and older people. There are no jobs. One works elsewhere—during the week, or for months at a time—or takes relief. The juvenile delinquency rate of the county is the highest in the state. The educational dropout rate is high and climbing. The influence of the church is waning. Priests, many of them unfamiliar with either the language or the customs, stay for a short while and are transferred elsewhere. Family ties are strong, but inter-family animosity is strong too, and of long duration. Those familes not on relief resent those who are. There is no cooperation between the towns, but a nasty, petty rivalry. No streets off the highways are paved; few are even gravelled. Electricity is available and used; phone service is available but little used. Each town has a useable domestic water system, but each needs major reworking, and Punta constantly finds its well running dry. The four towns support a total of two small general stores, two gas pumps, and two bars—both in Torreon.

Tajique, Torreon, Manzano, and Punta de Aqua are dying villages. They are products of a way of life that has gone and will not return. They are similar in their poverty and decline to hundreds of villages all over the United States. They are different from most in that they possess a sapped but still distinct culture, a life all their own. The people who live there, and the men who work far away in order to keep their families there, care about these towns. So do the outsiders who have come to know them.

Suppose, then, that we do care about Tajique and Torreon and Manzano and Punta—care too about the people who live in them. Suppose that we care about other dying villages and their inhabitants—villages all over the country. Can we really reverse their decades-long decay? It depends on how much we care. It also depends on *why* we care. American folklore still assigns mysterious virtues to the rural life—virtues that the city supposedly corrupts. Richard Nixon's speeches and the Republican platform's concern with the "problems of rural America" are twentieth century echoes of Jefferson's respect for the freeholding farmer and his distrust of the urban proletariat. A large part of our concern for farm and small town America is just nostalgia: a nostalgia felt not only by small-town expatriates but by life-long city-dwellers as well, a nostalgia for a too-placid world which never was. Today that nostalgia is enforced by feelings of guilt over the plight of the poor and the bypassed and the powerless—the miners and farmers of Appalachia, the sharecroppers of the black belt, the Hispano villagers of the Southwest. Neither nostalgia nor guilt, however, will command the will and resources required to revitalize our farm and small-town landscape.

Beyond all the self-conscious lamentation over the passing of rural America, however, beyond the shallow romancing over a time that never was, lies a real awareness of some unique values of small-town life—certain relationships among people, between man and the land. These values are not better than those of city or suburb, they are simply *different*. They are values worthy of respect and preservation—values that some people would like to share today. The villages and small towns of America are not dwindling and disappearing because their values are no longer meaningful, but because they no longer work economically, no longer provide the level of services and amenities that most of us demand. No dramatic violence is being done to rural America. It is withering away because it has little function in modern life. The question of whether it can be brought back to health is at base a question of whether it can once again be brought into the mainstream of American life, of whether it can be given a meaningful function. And if it can, there still remains the question of whether the cost would be worthwhile.

There are, in fact, two good reasons for not letting small-town America die, for trying instead, to revitalize it. The first reason is the possibility of radical changes developing within our society. Much has been written recently about the likely development of a true "postindustrial society." Much of it is wild and flimsy speculation. But from all this speculation there is emerging a sense that the world's more prosperous countries are likely to undergo a radical reshaping within the next several decades—a comprehensive social change perhaps equivalent to that which accompanied the Industrial Revolution. There is some agreement on the likely attributes of that changed society. Herman Kahn and Anthony Wiener (*The Year 2000*, Macmillan, 1967) have ticked off fifteen features of postindustrial society. Among them are a greatly increased per capita income with an effective floor on income and welfare, the diminishing roles of business as a major source of innovation and of the market compared to the public sector, widespread cybernation, the dominance of tertiary and quartenary rather than primary or secondary economic activities, the lessening importance of efficiency, a learning society possessing radically improved educational techniques and institutions, and the erosion among a large part of the population of traditional values oriented towards work, achievement and advancement. Some sociologists and physical planners have speculated in more detail on possible social and settlement patterns. Not all of their pictures are as gloomy as those of Huxley and Orwell. William Weismantel, for example, describes a possible postindustrial "city" characterized by radically new freedoms.

The postindustrial city is the spatial arrangement of
an economy which has successively seen its agricultural,
manufacturing and more lately its service sector auto-
mated. This automation frees people to leave the big
city just as automation of agriculture freed people to
leave the farms. City people are free to return to the
farms, stay in the industrial city as spectators, or form
postindustrial cities. The individual is primarily a consumer rather than producer. Free of the demands of
mass-production and the logical communication
needed to conduct industry and commerce in the style
of the industrial city, residents of the postindustrial
city can relapse to some of the more comfortable features of the preindustrial city such as kinship, concu-
bines, handicrafts, ritual and parochialism.

In a postindustrial society of spectacular technological sophistication,
the basic level of services available in the hinterland should approach that
of the future urban complex, and probably surpass that of today's cities.
In a society where people are largely free to choose where they live, the
opportunity to enjoy different amenities, different life styles, will largely
determine the choice made. Many people, at such a time, will choose to
live in small towns, or in the open country, simply because that is the life
they prefer. If all this seems like far-fetched speculation, consider the
increasing number of people—the wealthy, the retired, the artists and
writers, the hippies, who are making that choice today, in the face of
handicaps which in the future will not exist.

The second reason for concerning ourselves with rural and small-town
America is more immediate. It rests not upon speculation about the future, but upon the realities of life today in our cities. We have spent decades ignoring the warnings of those who have told us that our cities were
getting out of hand. We have looked upon Mumford and others as quaint,
slightly old-fashioned moralists who were out of tune with our world,
gloomy prophets who could not, or would not, comprehend the realities
of a dynamic, developing society. [But in 1969], suddenly, three sum-

mers of discontent have placed their warnings in a frighteningly contemporary context. Our most important cities have grown so large, their workings so complex, that many are choking on their own entrails. They have become great unplanned machines, systems more and more susceptible to progressive degeneration at the malfunction of minor parts or subsystems, whether that malfunction be a garbage strike or a ghetto riot. Planners have never thought that big meant better, but most have accepted increasing size as a prerequisite of the unquestioned benefits of urbanization. But it is becoming clear that size *does* matter. Sheer size is not the only handicap the city faces, but it is an important one. It will be a long time before the city goes away. In the meantime, arresting its growth and siphoning off new and existing population pressure, while not the only answer to our urban ills, is at least a possible part answer.

It is an answer that planners are finally beginning to consider. They speak of moving residents from the central city to the suburbs—remodelled suburbs, of course, suburbs more dense and diversified, suburbs willing to accept blacks and blue-collars, but still suburbs within a metropolitan complex. Others speak of new towns: large new communities of over one hundred thousand inhabitants, exemplary products of the planners' order and the designers' art, self-contained communities sitting in a pastoral landscape near existing cities. But almost no professionals have asked whether industrialization and technology might have progressed so far that we could settle people much farther from the great urban centers, or whether the effort and money required to restructure the suburbs or start new towns might just as well be spent in a national effort to revitalize selected small towns and villages—an effort that might cure some problems of *both* city and countryside. The idea of working with smaller, existing settlements need not replace the other more currently fashionable approaches. It could exist alongside them perfectly well. It needs only be one part of an overall redevelopment plan. It would seem an obvious approach, one well worth trying. Sadly, few are suggesting it. Only the voice of the politician is heard, seeking votes from his bewildered rural constituency.

Such a dual attack on the problems of city and village will require more than enticing an occasional small industry to some small town or other. It demands a new planning framework. This new framework must start not with the urban center but with the region. Every enlightened planner will say that he works with regional patterns. In practice he works within regularly defined areas. His power and his influence operate within political boundaries that often bear little relation to the overall settlement patterns of the urban complex and almost never consider the larger regional landscape. The scale of planning lags far behind the growing human reach. Where regional planning exists, it is oriented to the urban center. The countryside is to be a green visual relief for the harried commuter or the weekend refugee. There are exceptions. Some are as old as the New York Regional Planning Association or the Tennessee Valley Authority (TVA). A few states have planning offices that attempt to work seriously on this scale. The name of the game, though, is still "city planning," and it matters little that "urban" is substituted for "city," or that "and regional" is tacked on as an afterthought. The planner works in the city; the countryside fends for itself. The city is where the people are, to be sure, but there are people in the countryside too. Tomorrow there might be a good deal more of them. No scheme of regional planning will succeed until it grants that the small town is potentially as useful as the city, and is worthy of the same respect and concern.

The new planning, then, demands that the region, not the city, be given first consideration. Its second demand is that all settlements in the region be considered as parts of a network, an interconnected whole. Revitalizing small-town America cannot mean returning to the old self-sufficient small town. That town is gone forever. Just as we accept the fact that people now live in the suburbs and commute to the city to work, we should accept the fact that people could live in a small town and work somewhere else. An industry in every town is a dream. Several industries, in one or more locations, drawing their workers from many villages and towns, is a prospect that we can achieve. If and when a postindustrial society finally evolves, people might still choose to live in one town, shop in another, drink in a third.

Within this network, services could be located for their social benefits, not just for their economic efficiency. In a postindustrial landscape, educational and medical and other public-sector services might be much more important determinants of the network than jobs. Each service could follow its own hierarchy; each hierarchy could have its own unique spatial pattern. The secondary school might be in one town, the clinic in another, the sheriff's office in a third. If service distribution could be a sociological function, so could population concentration. People would settle in the larger towns not because they had to economically, but because they preferred more diverse and varied relations with other people.

Lastly, the new planning must accept the one factor that would make much of this regional network possible—that planner's abomination, the automobile. The private car might be choking our older cities; to the countryside it offers a wholly new freedom. In a landscape of open country and small towns and decent roads, fifteen miles can be driven in twenty minutes. A fifteen-mile radius produces an easily accessible convenience domain of over seven hundred square miles. Planning for development in now sparsely settled areas would take this convenience domain as its basic module. Services and convenience facilities would be placed into a network of these domains in such a way that a school, decent shopping, and a place to work would all be within fifteen miles and twenty minutes of one's house. It could be as simple as that—if we chose to make it so.

These are the functional aspects of a new approach to planning. But any new settlement pattern on such a scale must be more than just functional. We have always assumed, until recently at least, that our cities were clean, safe, and reasonably convenient. Generations of urban planners from Daniel Burnham to Jane Jacobs have asked for more than that. They have asked that the city be beautiful. In the last decade they have cried for more density and variety. We now ask that the city offer diversity of experience and richness of character. We ask that each city have its own special character. We ask that within the city each neighborhood have its own sense of place, where a man can say, "Here I am, here I live" —even in the ghetto. We should ask no less from the countryside. Each new or renewed settlement should have its own character. Each hamlet or village or town should be a *place,* its *own* place. This is not a matter of fake historicism or artsy-craftsy architecture. It is a matter of respect for things existing, subtle patterns of place woven from vistas and street widths and the siting and color and scale of stores, houses, and trees. The automobile could be the liberator of the landscape. Too often it is becoming the great leveller. It is not only making one place look like the next, it is running one place into the next. It is creating its own rural sprawl, a sprawl of little houses stretching along both sides of every decent road out of town, extending many times the length of old Main Street. If the countryside is to prosper, it must be different from city or suburb. Not better, for that is a matter of individual taste, but different. That difference is in part the simple business of containing our towns and giving them boundaries.

To see how these principles could be applied, it would be useful to look at areas where a new planning framework would work the most radical changes—regions with few people, where present poverty and decay exist in a landscape of undeveloped richness. The depressed mining country of West Virginia and Kentucky is one such region which is already the subject of national publicity and concern. Nowhere, however, is the potential for development more dramatic than in the high, arid, sparsely settled lands of the American Southwest. Perhaps Tajique, Torreon, Manzano, and Punta would be good places to start. A new approach to planning is indeed the only hope that these villages have. Without it they will die, probably within a decade. The villages lie in a setting of fine climate and ample space, of natural beauty and recreational potential, and of rich historical and cultural interest. These are amenities that are highly prized now and will be even more valuable in attracting people in the future. But if the landscape is beautiful and varied it is poor and empty. In our concern with megalopolis we forget that over large areas of the country the major problem is not too many people, but too few—too few people to support adequate jobs, decent water and sewer systems, diverse social intercourse. This land needs more people. A new regional planning could supply them.

Those amenities mentioned are common to many parts of rural America, as common as poverty and emptiness. But the landscape of the Southwest has two unique amenities that could be the primary determinants of new settlement pattern at the largest regional scale. One is an immensely rich and varied ecology. New Mexico, in its wide altitudinal range, contains six of Merriam's life-zones. The four hill villages share not only three of those zones in their immediate surroundings, but share the differing plant and animal life of both the Great Plains and the Inter-mountain West. The other unique feature of the Southwestern landscape is its vastness. It is land of awesome and overpowering scale, of distance and openness defined and exaggerated by sharp relief and changing color and shadow, a land where telephone poles and barbed wire fences and billboards are not eyesores, but welcome assurances that somewhere there are people. But if its ecology and its loneliness are unique, they are all too fragile. A handful of people in the wrong places will destroy both —quickly and permanently.

There, of course, is the problem. How do we introduce new settlers with all their paraphernalia of affluence into the landscape without de-stroying those very amenities which they came to enjoy? The answer is simple. Give people plenty of room or cluster them together. It is a mat-ter of permitting only extremes of density—one family per hundred or more acres, or six to ten families per acre—that is, the density of the farm or small ranch, or the density of the village. In between these extremes there is only the ruin of the land.

If the great regional landscape has its own features, so does the smaller landscape surrounding the villages. Driving south along Highway Ten is like driving along a seashore. To the east, the range land rolls off flat to the horizon, interrupted only with hills rising like small scattered islands, and ranch houses sitting alone like ships afloat. To the west, the wooded mountains send down fingers of green, and the villages sit at their fringes or between their promontories like ports in coves. The two density ex-tremes would work here in dramatically simple form. East of the road: open land, where even one family to one hundred acres would be too dense a settlement. While one house each sixth of a mile along the road is not exactly a road town, it is enough to spoil the character of the land-scape. The answer is an even lower density—one approaching the historic ranch pattern. One house per full section is what we need. If this is eco-nomically impossible, then permit six houses per section clustered in an area comparable to that occupied by the old ranch house and its out-buildings. Along the road and west of it: village-type density. Fill in the existing villages first. There are plenty of abandoned but renewable buildings and plenty of vacant sites. For more people, build more ham-lets—along the road or on fingers of land reaching up toward the moun-tains. There is no need to search for new patterns of settlement—what exists is right.

Within the immediate landscape, each settlement, no matter how tiny, should have its own character, too. Each should be a place. For the scat-tered buildings east of the road, fences and trees should be just enough to stake out a personal domain on that flat large plain. In each village the job is to discover the sense of place and preserve it. Luckily, the limited local building skills and the prevalence of adobe mud and local stone still make the traditional building methods the cheapest. The existing street patterns and house groupings should be respected and reinforced, or old ones renewed. Each village has its own pattern. Torreon, though con-fined, is still a road-oriented town. The empty land facing the church could be used for two or three stores, and a small plaza could be shaped. This would supplement the linear pattern of the town by adding variety. Manzano was originally built around a plaza. Now Highway Ten cuts through the plaza. It is to be widened in a few years. Move it only one hundred yards to the east, to the edge of the town, and the plaza is again a focus. For years, the ranches and villages have been characterized by two kinds of trees—cottonwoods and poplars. Their quaking green leaves, bright against the silvery grass or the dark conifers, are a clear sign of where people live. New landscaping could do no better. New hamlets can find their own form. Some should be for people who like the big sky, who like to look out over the emptiness of the basin. Others should be for people who prefer the intimate enclosure and shelter of the wooded foothills.

Above all, the roadside between the villages should be kept free of buildings. The bare roadside does more than emphasize the meeting of hills and plain and display the scale of the landscape. It makes each vil-lage an event, a place distinct and separate from its neighbors and from the open land between. This is true not just of these Spanish villages, or of the Southwest. Of all the measures that could be taken today to improve the economic health and visual appearance of rural America, none would be more helpful than the simple business of confining rural commercial use within the limits of the existing towns. The strip has its place, but that place is not the open countryside.

The traditional patterns seem completely right, needing only renewal and intensification. Why? Neither the Spanish villager nor the Anglo rancher has ever been a great respecter of the land. He has eroded it, overgrazed it, and littered it with trash. But at least his buildings have never sprawled senselessly all over it. Perhaps some visual feeling for the landscape played a part in this, but probably even that was missing. The Hispanos clustered together because of their land-holding customs and for defensive reasons. Later, the Anglo settled upon a section or more because less would not support a family in a market. But they settled just right. Everything fits. If we do begin to resettle rural America, we will often find that same sense of fit, that same harmony with the land. For it is not a wilderness that we will be moving into, but a working landscape, that has felt man's hand for many years, a landscape that has had rules. It might not be as beautiful as the land where the Manzanos meet the high plains, but almost always it will be reasoned, ordered, and coherent. The determinants that shaped that landscape have declined in importance. It will be a long time, however, before a postindustrial society develops equally clear determinants. Until such new determinants do develop, we can do no better than to restore and revitalize existing patterns.

How would we start on a job of such enormity? Where would we begin? With the will and the money, of course, and with new tax structures, and with the reshaping of the myriads of channels of political authority and planning responsibility that bear no relation to the realities of the regional landscape, and with a hundred other things. We start by considering the needs of the people who live in that countryside now, people who cannot afford to wait upon a postindustrial society, people who need jobs and better schooling and better medical care today. Because a postindustrial society has not yet arrived, we start near cities and major highways; we start, regretfully but necessarily, with tourism and recreation and summer cabins and commuter houses and the like. But always we must remember that the long-term future of the countryside will depend upon its ability to attract people away from city and suburb. In a society of cybernation and leisure and biological engineering this will require jobs that can survive automation, and educational and medical services that are not just adequate, but good, truly good.

These are the abstract problems, the mechanical problems. The biggest problem may turn out to be people. And if Punta and Manzano, Torreon and Tajique, are all rich in the beauty of their land, they are rich too, with unresolved social conflicts. One of these conflicts is the familiar one of rural-urban political conflict. We have been told for a decade that the political power of an agrarian minority has thwarted progress towards solving our urban needs. Perhaps that minority has had too much power yet too little—enough power to frustrate the cities, but not enough to re-structure the countryside. Certainly, the immediate future of the Tor-rance County hill villages lies in catering to the needs of an expanding Albuquerque and its Bernalillo County. Yet the government of Torrance County is reluctant to join any inter-county body that includes Bernalillo County; it prefers to ally itself with the rural counties of eastern New Mexico. This attitude represents more than backwardness, or provincial-ism. It represents the fear of being swallowed up in an urban system which displays no understanding of, or respect for, rural areas.

The attitude of the Hispano towards the Anglo is another such con-flict. The Anglo summer resident might be an economic asset, but the villagers seldom welcome him. For too long, the Spanish-American has done the Anglo's menial work. The villagers are now poor, but they live on their own land amidst their own friends and family in their town. Over the years they have seen come and go the politicians, the Ph.D. stu-dents, the county agents, and the summer VISTA (Volunteers in Service to America) workers. These people, in the villagers' eyes, have promised much and delivered little.

A few years ago the schools in the villages were closed. The students from the villages are now bussed into the Anglo towns of Estancia or Mountainair. The villagers claim that the dropout rate was almost nil before this happened. This might or might not be true. What matters is that the villagers believe it. The centralization of schools perhaps improved the economic efficiency and the academic program. It also removed the last village function of any relevance to mainstream society. For at least the last two decades, many of the outsiders working or settling in the mountains east of Albuquerque have had some understanding of local problems and an empathy for the area's history and culture. They have fared poorly enough with the villagers. Whether a much larger influx of real "foreigners" would produce more open resentment and active hostility, or perhaps even hasten the death of the existing communities, is hard to tell. It is, at any rate, a real danger.

Many of the area's residents, Anglo and Hispano alike, still live with a frontier spirit. That spirit finds its outlet not in homesteading or mining, but in the desire for a trailer on acres of land, or a house built with one's own work over the years, with no building codes or zoning regulations to fool with. It is a spirit that may, in the end, be incompatible with clustered settlements and ecological planning. If it disappears, or if it is crushed out, even for the best reasons, it will be a loss. After two decades of grappling with the complexities of urban civilization, we are learning, painfully, that the "big" problems of money and land ownership and political authority often turn out to be less important than the attitudes of people.

There are no easy answers to these problems. The absence of any existing long-range rural development plans could turn out to be an advantage, however, for the communities' awareness of their own problems and wants is preceding planning. In New Mexico VISTA is beginning to recruit its summer associates not from among out-of-state college students, but from the small villages within the state. In Torrance County, the Community Action Program, after little success with such ventures as village libraries and food-buying cooperatives, is sponsoring candidates in village and county elections. It might be that the initiative for planning in rural America will come not from the planners' offices but from the community. Advocacy planning might become the foundation of the planning process and not a belated reaction to it. Such a development might avoid much of the grief to which urban planning has so lately come.

There is, then, a kind of planning that could restore the withered vitality of Tajique, Torreon, Manzano and Punta, a kind of planning which might help solve some of our urban problems as well. It is a kind of planning that could offer another distinct way of life, an alternative to the endless suburbanization of our landscape. It offers a new domain of experience, and it promises a unique sense of place. Such an effort seems most unlikely now. Only ten years ago, the current concern for racial justice would have seemed equally unlikely within our lifetime. Perhaps the accelerating processes of urban disruption and rural decay will force a radically new concept of planning upon us sooner than we think. With luck it might not come too late.

PERCEPTION AND EXPERIENCE

The biggest difference of all was to the ear; you then traveled through the sounds of the country quite as much as through its sights.
—Edgar Anderson

Appearances lead people in all cultures and eras to assume that some things are 'man made,' others natural, when the truth would reveal the opposite, or more likely an intermingling of agencies. —David Lowenthal

What makes one landscape appear harmonious, another incongrous, is the entire experience of the viewer. —David Lowenthal

An abstract nature, as it were; a nature shorn of its gentler, more human traits, of all memory and sentiment. The new landscape, seen at a rapid, sometimes even terrifying pace, is composed of rushing air, shifting lights, clouds, waves, a constantly moving, changing horizon, a constantly changing surface beneath the ski, the wheel, the rudder, the wing. The view is no longer static, it is a revolving, uninterrupted panorama of 360 degrees. In short the traditional perspective, the traditional way of seeing and experiencing the world is abandoned; in its stead we become active participants, the shifting focus of a moving abstract world; our nerves and muscles are all of them brought into play. To the perceptive individual there can be an almost mystical quality to the experience; his identity seems for the moment to be transmuted. —J.B. Jackson

HORSE-AND-BUGGY COUNTRYSIDE

Edgar Anderson

It is gone and gone forever; like youth it was not fully appreciated until
it was over. I do not wish it back; it was for me an inconvenient and a
confining landscape, but neither should it be forgotten. Protest wells up
that its sights and sounds and smells are lost forever. That it should go
without a record of what it was like, some indication of what it meant to
those who lived in it, seems as out of place as the passing of a great
statesman without a word as to his services to the country. Miss Cather
has written very beautifully about the dust of country roads; there are
phrases here and there in Hamlin Garland; but I do not remember having
found anywhere an attempt to record the impact of the horse-and-buggy
landscape upon those who lived and moved in it.

Impact it had and a very sharp and constant one. You couldn't then
be unconscious of the terrain over which you were traveling. The road re-
curringly thrust itself upon your senses; there were vibrations, sounds,
and smells. The wheels commented almost constantly about the roadway.
There was a soft even purr as they pulled slowly through the sand, an-
guished crunches in sliding over the edges of big stones, a taut, almost
musical vibration when they whirled rapidly along on good gravel. There
were waits for the horses to rest when the grade was steep. Swamp roads
were an adventure; plank bridges over the ditches, heavy new gravel that
almost stalled the wheels, miry spots where the buggy lurched unevenly
and, sometimes, bumpety-bump stretches of corduroy when the logs

[from *Landscape*, vol. 4, no. 3, Spring 1955]

and brush which held the road up out of the swamp were still close to the surface.

Such roads had their own standards of courtesy. I remember the story told me by the first man to drive an automobile on the back roads around Fairlee, Vermont. He came up behind a farmer with a load of hay midway through the balsam swamp. It was a narrow road; wagons could just barely pass if they met head on, and a faster team, coming up behind a slower one, stayed behind until the swamp had been crossed. My friend tooted his horn vigorously, but there was no response from the hay wagon. Finally he stopped, stuck out his head and yelled, "Hey, don't you hear me up there?" The farmer reined in his horses, stood up, looked down with great dignity and said, "Oh yes, I hear you all right," and drove ahead slowly and unconcernedly to the end of the swamp where he pulled off to the side as he would have for any other vehicle which had come up behind him. "He was right and I was wrong," said my friend years later, "but I didn't realize that at the time."

In those days differences in vegetation between a main highway and a side road were even greater than now. There weren't many state highways but the best of them stretched straight ahead with mathematical perfection, even though just of gravel. These rights-of-way were usually wide and from the road edge to the boundary fences were mowed pretty much as now, though road building had not been preceded by earth moving. As a result, there were more native plants in among the grass and fewer of those cosmopolitan tramps which take so readily to disturbed habitats. But the side roads, those fascinating side roads, they are mostly gone and there is nothing in the modern road system to compare with them. The wheel track wound here and there, depending upon the slope and the vegetation and the character of the land. Sometimes it was straight for a little way; frequently it wobbled. There was often grass in the roadway outside the actual wheel tracks; shrubs like sumac and elderberry pressed so close to the road that you could smell them as you drove by and children snatched at the flowers. Accommodating drivers of the local stage-line learned to snip off small twigs with a snap of the buggy whip and present them to lady passengers.

The biggest difference of all was to the ear; you then traveled through the sounds of the country quite as much as through its sights. On clear days in spring there was the incessant calling of the meadow lark, and in the heavy heat of early summer you heard repeatedly the scolding call of the Maryland yellowthroat coming out of the dusty bushes along the roadside.

Even more characteristic than these sharp sounds were the soft and gentle ones that are scarcely heard any more, anywhere. So crisscrossed is our modern landscape with the pounding of Diesel trucks and farm tractors that over square miles of countryside, even when standing quietly in fields or woods, there is enough interfering of mechanical noise to keep one from hearing the softest of the cricket trillings on autumn evenings, the gentlest of breezes in the long-needled pines, or that magic sound of the wind in the young corn. When the corn is about waist-high, the leaves are not yet at all stiff. They bend and twist in the wind and touch each other lightly. If there are no other noises, you can hear the slightest breezes rippling here and there through the cornfield. There is no other sound like it; on a clear, quiet day in early summer when the wind rises and then dies away again, this mystical rustling (like thousands of silk petticoats just barely being stirred to sound by gentle movements) scurries back and forth over the landscape.

Or the onset of snow; who hears this now? The first flakes as they start to fall hit the dry leaves and the dead grasses along the roadsides with tiny, sharp noises, like the ghosts of mice in baskets of vanished waste paper. They begin as sharp, separate rustles, becoming more and more blended as the snow falls in earnest, cushioning the cart wheels into silence. It was beautiful to hear, but to the alert traveler it was a sign of trouble ahead, and if one had far to go, it had ominous associations.

No, the horse-and-buggy landscape was not just delight. The finest days were finer; one got everything from them that could be collected from a fine day, but the harsh ones were harsher. There was misery as well. I remember the farm families driving home in the early dusk of winter afternoons. I used to meet them in the swamp road, as I trudged home after a Saturday tramp through marshlands frozen so solid that one could easily explore the quagmires which were impossible in the summertime. The farmers would have done their weekly shopping and the wagons jolted noisily over the frozen ruts. The driver usually sat alone up front, hunched down against the bitter wind. His wife and children, together with the purchases for the week, were down in the straw that had been spread over the wagon bed, swathed in heavy horse blankets and old lap robes. Jolt! Jolt! Jolt! Jolt! Every step of the way told them of the bitter cold, the frozen road, the strain on the tired horses; for most of the families I used to meet there was at least another hour yet to be faced.

NOT EVERY PROSPECT PLEASES

David Lowenthal

I begin with a digression. Why do people feel as they do about particular landscapes? My interest in this question was first aroused by a surfeit of sermons on the virtues of the wilderness. However, notions about wilderness serve me principally as points of departure for a broader inquiry: what is it that makes any landscape seem pleasant or repellent, fitting or unsuitable, to people of various backgrounds and inclinations?

The Wilderness Cult

Early man, according to some of his latter-day descendants, viewed his physical environment with awe and dread. To him, Nature was a tyrant; he survived only by strict obedience to her dictates. To us, it is not Nature but our own creations that are terrifying. Nature seldom threatens most urban Americans; indeed, she scarcely causes us discomfort.

Nevertheless, a new religion of nature is in the making. Worshipers of nature exhort us from the pulpits of countless conservation societies and Audubon clubs; the President's Advisory Commission on Outdoor Recreation transmutes their dogma into national policy; *Life* magazine gives it a four-color imprimatur. Nature is wonderful, they tell us; pay homage to it in the Wilderness, and where there is no Wilderness, create Open Spaces.

The tone of Senator Wayne Morse's comment on the Wilderness Bill, in 1961, exemplifies the underlying rationale: "You cannot associate with the grandeur of this great heritage which God Almighty has given

[from *Landscape,* vol. 12, no. 2, Winter 1962–63]

the American people and not come out of such a trip a better man or woman for having come that close to the spirit of the Creator." Americans are near a state of grace, runs this argument, thanks to their past and present intimacy with the wilderness. Our forefathers mastered a continent; today we celebrate the virtues of the vanquished foe: To love nature is regarded as uniquely American. "The outdoors lies deep in American tradition." chants *Outdoor Recreation for America.* "It has had immeasurable impact on the Nation's character and on those who made its history. This is a civilization painfully and only recently carved in conflict with the forces of nature. . . . When an American looks for the meaning of his past, he seeks it not in ancient ruins, but more likely in mountains and forests, by a river, or at the edge of the sea."

But these are pious sentiments, not historical facts. Our civilization is mainly imported from Europe, and has had more to contend against in Main Street than in the forests and mountains; pioneer life was usually duller and safer than it is generally imagined; and Americans who seek the meaning of their past look for it in Europe, or in Bunker Hill, Gettysburg, Williamsburg, Freedomland, and the deserted mining towns of the Sierras, precisely because such places are humanized, not wild.

Converts to the wilderness cult, like conservationists generally, tend to feel that their preferences are not only more virtuous than those of others but also more enduring. "Conservation should preserve all the fine things in life. Our heritage must be preserved for those who come after us." —so runs a typical admonition. Long-term or eternal goals are considered ethically superior to immediate ones; and future generations are expected to accept present-day precepts about what is important. Yet our great-great-grandchildren may care little for our image of the wilderness. Moreover, a heritage cannot be preserved intact; if it were, it would have only antiquarian interest. Tomorrow's patrimony is bound to be different from today's. The wilderness was no 'heritage' to folk who had to cope with it; it became one only when it no longer had to be lived in. The same is true of pastoral landscapes, rural villages, even of 19th Century industrial centers; and landscapes now despised will some day be prized as a precious heritage.

The wilderness is not, in fact, a type of landscape at all, but a congeries of feelings about man and nature of varying import to different epochs, cultures and individuals. For Elizabethans, the wilderness was barren, chaotic, frightful, "howling"; for contemporary Europeans, it is often associated with primitive and romantic tribes in distant lands; for many Americans, it is an entity distinct from the workaday world, an oasis where the laws of nature still apply.

The cult of wilderness—its origins and history, its philosophies and programs, its impact on landscapes and its implications for mankind—should receive thorough scrutiny. The way Americans feel about wild nature is intimately bound up, as has been suggested, with this country's special history. But the wilderness is only a part of anybody's picture. And attitudes toward wilderness can hardly be understood except in the context of beliefs and assumptions concerning nature and landscape generally.

Elsewhere I have examined some causes and consequences of supposing nature to be good, wise, purposeful and balanced. ("The American Image of Nature as Virtue," *Landscape,* Winter, 1959–60). Let me discuss some assumptions about the *appearance* of landscape—assumptions concerning its origins or antiquity, its cultural context or function, and its esthetic congruity. For judgments about the way things ought to look are conditioned by how things are thought to have come into being, how old they are, what they are used for, and how well they fit into their surroundings.

The Genetic Vision

Comfortably ensconced in a Pullman car moving through Pennsylvania, the book critic J. Donald Adams saw something from the window "which revolted me, and . . . made me boil with anger. We had been passing through hilly, forested country, still lovely, even in its second growth," writes Adams in his weekly *New York Times* column, "when suddenly I saw a mountain stream so discolored, so noxious as almost to turn one's stomach. It was easy to imagine how it had once looked, sparkling and clear, before the mines somewhere beyond had polluted its waters and made it a thing of blasphemy."

What seems ugly to Adams he condemns as evil; but which of these judgements determined the other? He cannot be certain that the stream is polluted—he is no hydrologist. Suppose he learned the water was discolored not by mining but by a sulfur vent upstream, or simply by turbid spring floods; would he still find it 'noxious'? If he knew nature were responsible for the color of the stream, would he "boil with anger"? Would he still conjure up an image of pristine purity? In short, Adams takes it for granted that landscapes he enjoys are natural and that those he dislikes have been ruined by man. Nature beautifies; man deforms. To alter the natural order is inherently wicked and *hence* displeasing to the eye. The forest, like the stream, is judged by this dictum. "Still lovely, *even* in its second growth," the wooded hillside is more attractive than some denuded slope, but lacks the perfection of the primitive forest.

Where landscapes are not patently marred by man, it is all too easy to assume that they are primeval. To the city dweller, anything that is not geometrically arranged may seem in the realm of untouched nature. He seldom distinguishes the pastoral from the wild. Indeed, in a landscape seen for the first time every component—mountains, old houses, even billboards—is apt to strike the viewer as a durable fixture of the scene.

But no matter how natural or untouched they may look, most landscapes have been profoundly altered by man, directly or indirectly, over long periods of time. "A field of pasture grass looks as if it had been there forever," remarks Edward Hyams, "and it is hard to absorb the fact that, like any urban square, it has been imposed on the wild." One intuitively feels that if man had interfered with the lovely vista it would bear obvious marks of manufacture, alteration or decay. "Surely a water-meadow, with its lush grass kneehigh to the cows and its decoration of tall golden buttercups . . . is a gift of God, or nature." Not so; "we owe it all to the greed for gain of a speculating 17th Century nobleman and the engineering skill of some Dutch drainage expert." On the other hand, some aspects of nature seem artificial. Vegetation is apt to look more 'natural' than barren rock, because we perceive analogies between geological and architectural structures—many tourists, asserts Joseph Wood Krutch, assume the Grand Canyon is at least partly manmade.

Misconceptions about man's role in landscape formation are universal and probably inevitable. Appearances lead people in all cultures and eras to assume that some things are 'man-made,' others natural, when the truth would reveal the opposite, or more likely an intermingling of agencies. What is unusual—and, to my mind, destructive—is to combine wrong notions of lanscape genesis with the moral judgment that men generally act for mean motives and with effects offensive to the sight.

The contemporary tendency to find beauty and good in the 'natural,' ugliness or squalor in what man dominates, is not only moralistic; it is an esthetic aberration in the history of landscape taste. The contrary view has usually prevailed. Thus to Daniel Defoe, Westmoreland was "the wildest, most barren and frightful" county in England; he liked nothing about it except "some pleasant manufacturing towns." In most canons of landscape beauty, man and his works occupy a prominent place. A century and a half ago, in a fairly representative view, Arthur Young described the clang and smoke and flames of the iron forges at Colebrook Dale as "altogether sublime." Russians today similarly admire Kochegarka mine, the "pride" of Gorlovka, with its magnificent slag heaps scattered over the steppe for miles around.

By and large, men find lived-in landscapes more attractive than wild ones. "We have in places made the Earth more beautiful than it was before we came." Sir Francis Younghusband maintained in his Royal Geographical Society address: "I can realize what the river-valleys of England must have been like before the arrival of man—beautiful, certainly; but not *so* beautiful as now. . . . Now the marshes are drained and turned into golden meadows. The woods are cleared in part and well-kept parks take their place. . . . And homes are built . . . which in the setting of trees and lawns and gardens add unquestionably to the natural beauty of the land."

Raw nature was condemned also by 19th Century Americans. The Rockies and the Sierras displeased early tourists. "The dreariness of the desolate peak itself scarcely dissipates the dismal spell," one traveler wrote of Pike's Peak, "for you stand in a hopeless confusion of dull stones piled upon each other in odious ugliness." Americans preferred the picturesque—caverns evocative of cathedrals, pillars resembling ruined castles. If they could not have real ruins, they wanted make-believe ones; they had had a plethora of wilderness. To admire, much less love wild nature, as Daniel Beard advocated, initially seemed ludicrous to almost everyone. Indeed, the first forest reserves were called "primitive areas" because, according to Earl Pomeroy, the government feared "the public might find the word 'wilderness' repulsive."

All that is changed; for contemporary Americans, it is civilization that is hard to endure. A 'primitive area' is now a place, according to the Outdoor Recreation Resources Review Commission, in which one enjoys a 'wilderness experience'—"a sense of being so far removed from the sights and sounds of civilization that he is alone with nature." The implications of this statement underlie most of the issues I have raised. Are the "sights and sounds" of civilization more difficult to live with than those of nature? Is being alone with nature the only, or even the best, alternative to civilization? What about being alone in rural or pastoral countrysides, or in landscapes full of historic ruins but void of present day activity? And how much do these distinctions matter if one is not aware of them? Consider an urbanite walking in woods where human enterprise is apparent to any Indian or woodsman, but invisible to him; does he not feel as alone with nature as he would in a truly primeval landscape? Alternatively, if J. Donald Adams suspected the green on leaves came from manufactured chlorophyll, would he not find the forest ugly? Oranges look unnatural only when we are told their color is artificial.

Few Americans, however, are devotees exclusively of the aboriginal; many are fond of the merely old. Landscapes that cannot qualify as pure wilderness may still be acceptable if they are sufficiently ancient, uncontaminated by contemporary life, or relics of an epoch when (it is assumed) man lived in harmony with his surroundings. Americans tout no particular era as an acme of esthetic virtue; all they ask is certified antiquity, preferably numerically precise. It is the anachronistic appearance that is all-important. Critics of Malraux's clean-up program for the *monuments historiques* of Paris protest that rejuvenation "has robbed them of their weight of reality, that the Concorde palaces now look like Hollywood movie sets of the Concorde palaces." And the Hotel Crillon rejected cleaning outright because, noted Genêt in the *New Yorker*, "its clientele, especially the Americans, had faith in the dirt of the facade as a guarantee that they were living in a genuinely historic old place." Few people are really taken in by such a 'guarantee.' But this matters little, as we are willing to be fooled even when we know we are being fooled.

At home, visual self-deception is unabashed. Dirt is all right on European facades; here, it is matter out of place. For every restored Williamsburg, there are countless generic prototypes like Old Sturbridge Village—sanitary facsimiles rather than actual places, which rely for verisimilitude on scholarly precision. Self-conscious and sentimental about the past, we fence it off in special landscapes of its own, as if history were a zoo. Repair jobs for Niagara Falls and other 'natural' scenic wonders are deemed proper because they preserve or restore nature's creations; artifice is the means, not the end. But even where entirely new effects are produced, as with floodlights in caves or music piped out of scenery, the landscape passes for 'natural' as long as the face of the earth is not grossly and visibly altered. In the classic phrase of a Long Island developer, "We don't tamper with nature . . . we improve upon it!"

The Functional Vision

What Americans see in landscapes, and how well they like them, depends also on how the landscapes are being used. Form is supposed to fit function; camouflage is arty and dishonest. We like to call a spade a spade: however ugly a factory, a city dump, a used-car lot may seem to the passerby, as long as it fulfills its function it is presumed to look all right. If the used-car dealer didn't like the way his lot looked, he'd do something about it—not for his own visual pleasure, to be sure, but to attract customers. To enjoy a landscape or to commune with nature, we go to the wilderness. We do not prettify the rugged face of workaday America.

The landscape, in short, is worthy of its hire. Its ultimate critics are its residents, not its visitors, however unappreciative the former, however learned or perceptive the latter. This is the burden of a passage in William James's "On a Certain Sense of Blindness in Human Beings":

> Journeying in the mountains of North Carolina, I
> passed by a large number of 'coves,' . . . which had

been newly cleared and planted. The impression on my
mind was one of unmitigated squalor. The settler had
in every case cut down the more manageable trees, and
left their charred stumps standing. The larger trees he
had girdled and killed . . . and had set up a tall zigzag
rail fence around the scene of his havoc to keep the
pigs and cattle out. Finally, he had irregularly planted
the intervals between the stumps and trees with Indian
corn. . . . The forest had been destroyed; and what had
"improved" it out of existence was hideous, a sort of
ulcer, without a single element of artificial grace to
make up for the loss of Nature's beauty. . . . Talk about
going back to Nature! I said to myself, oppressed by
the dreariness. . . . No modern person ought to be willing to live a day in such a state of rudimentariness and
denudation.

Then a mountaineer told James, " 'Why, we ain't happy here, unless
we are getting one of these coves under cultivation.' "

I instantly felt that I had been losing the whole inward
significance of the situation. . . .To me the clearings
spoke of naught but denudation. . . . But, when *they*
looked on the hideous stumps, what they thought of
was personal victory. The chips, the girdled trees, and
the vile split rails spoke of honest sweat, persistent toil
and final reward. . . . In short, the clearing, which to
me was a mere ugly picture on the retina, was to them
a symbol redolent with moral memories and sang a
very paean of duty, struggle and success.

And he points the moral: "The spectator's judgment is sure to miss the
root of the matter, and to possess no truth."

James considers the mountaineers' impression truer and nobler than
his own, not because they see better, but because they have a better right
to judge. Beauty is in the eye of the beholder, to be sure; but which beholder is right? Why, the man who is usually on the spot—presumably
the shepherd on the Downs, the sexton at St. Paul's, the elevator opera-

tor in the Eiffel Tower. He may have no time, training, or inclination for an esthetic appreciation of landscape; no matter, he acquires it by osmosis.

The Esthetic Definition

Preoccupation with purpose is in fact no aid, but a deterrent to landscape appreciation. The man who has to consider what things are used for is least likely to note their shapes, colors and patterns. To exalt his judgement is to promote a complacent inattention to appearance, an abnegation of esthetic response. The visitor's view is discounted, but the resident is too preoccupied to have a view.

There are a few who care, or pretend to care, about the way things look where they live. But many homeowners are satisfied by linguistic landscaping. Through emotive names—even by such geographical fantasies as Hilldale Heights, Valmont Gardens, and Glencrest Homes—"an idealized landscape is imposed on an existing one by verbal *force majeure*," as Arthur Minton points out in *Names*. Maybe only God can make a tree, but any real-estate developer can create a veritable forest in the mind's eye of his clients.

Few people really look at the places they live in, work in, or travel through. Anesthetized against their surroundings, they spare themselves pain. "Start by looking at things, and then one becomes aware how hideous they really are," writes F. F. C. Curtis in the *Journal of the Royal Society of Arts.* "Some towns are hideous from one end to the other. I think it is because people do not open their eyes and do not recognize them as hideous."

The concept of scenic beauty, formerly tied to specific attributes, has since the 18th Century lost all precise meaning; it now promiscuously denotes anything that gives pleasure. "We talk about a 'sublime' or 'lovely' landscape," as Fenichel says, "because we feel sublime or lovely when seeing a landscape of this kind." No longer captives of the picturesque, we do not require that scenery look like a painting; and we can enjoy landscapes that point no meaning or moral.

Nonetheless, certain landscapes look right, others wrong or unhappily composed. They may seem incongruous because we have a pictorial bias against certain combinations of form and color, or a teleological outlook that condemns assemblages unless they look planned. Some of us find landscapes deformed unless the visual assemblage is familiar. Various objects, shapes or colors may seem out of place in certain settings. The totally strange and new may be less unsettling than a combination of known with unknown elements, or of forms belonging to two or more disparate cultural contexts. Those used to Western European or North American landscapes find something absurd and dissonant in lands where mules coexist with motorcars, and temperate zone flowers compete with tropical vegetation. The commingling of familiar and exotic, or of past and present, can be bizarre and disturbing.

Whatever the reason, most people expect a measure of visual harmony in landscapes. But the measure is personal as well as cultural: variety pleasing to one individual may pass unnoticed by a second and seem a shocking mélange to a third. Take litter. Most Americans pile junk in town dumps and automobile graveyards or throw it in rivers and down railway embankments. But streets and the countryside generally are free of cans and paper—otherwise, someone complains.

Other cultures treat litter differently. From an apartment in one tropical land I recently looked out at a garden, a stretch of grass that passed as lawn, and pleasant buildings opposite. What made the scene so tawdry? Garden and lawn were strewn with paper and cans. Litter collected there, months on end, because *no one saw it;* to the residents, cans and paper did not seem out of place among grass and flowers. And to many folk, here as well as abroad, cans and paper are not litter at all, but valuable materials for construction and fuel.

The problem is made evident and vivid when litter is mistaken for something view-worthy. Disillusioned by mistaking a piece of blue paper on a hillside for a flower, A. M. Sayers asks, in the *New Statesman, "Why* do we dislike litter? . . . Is the disgust rooted in fear of one's fellows as a whole? No. Ancient Roman or medieval 'litter' doesn't trouble us. . . . Is it to do with the Age of Paper? Why should amphorae be so much more respectable than old tin cans?" But some people see nothing disreputable in old tin cans. Everything depends on what is considered 'out of place,' on the rate of decay, on how rapidly artifacts cease to be junk and become part of the heritage. And these are matters of great complexity. Consider the reaction of Reyner Banham on Tair Carn Isaf, in Wales. Fascinated by the spectral, other-worldly quality of the cairns, the rocks and the light, he stumbles against "a stumpy concrete obelisk"—an Ordnance Survey triangulation point.

> And it jarred. . . . The triangulation-point just seemed to dirty things up. It destroyed the landscape as surely as the sheep that were eating the grass cover off the peat-beds. . . . The thing was a mess in itself, rust-stained from its metal fitments. It was a mess functionally, weathered down to a drab grey-brown that made it invisible against the landscape, and thus useless as a sighting-point. . . . Failed on looks, failed on function, failed for cultural inertia, too, for what else could force the classical obelisk form, apt to stone and primitive technology, on a machine-age material whose nature enjoins no form in particular?

Man's other contributions to that landscape seemed to Banham quite inoffensive—the ancient castle across the valley, his own aluminum-stemmed pipe and butane-fueled lighter. These belonged, visually, because they were or had been functionally effective. "Success or failure by the norms of the time," Banham concludes, is "almost an absolute standard for evaluating buildings in the natural landscape. For the landscape . . . can never be anything but up to date. Against the ruthless standard of nature . . . only the most spit-hot artifacts can survive. The trivial pipe is competitive, so is the portentous castle, the dirty obelisk is not."

The principle Banham enunciates is debatable. But even if it is accepted, landscape judgment would differ from person to person. Lacking his special knowledge of technology and architecture, most observers would probably not be bothered at all by the triangulation marker. The main point, however, is that all the considerations Banham mentions are pertinent. They do not merely justify esthetic judgment; they help to form it. What makes one landscape appear harmonious, another incongruous, is the entire experience of the viewer.

THE ABSTRACT WORLD
OF THE HOT-RODDER

J. B. Jackson

The long holiday-weekend approaches, and at quarter-hour intervals the
radio broadcasts the words of Mr. Ned Dearborn of the National Safety
Council predicting the total of highway deaths. Average Citizen listens
with a vague alarm. "Gosh!" he says, "436 deaths in a seventy-two-hour
period! *Gosh!"* And in the back of his mind he goes on planning the family's weekend trip. By leaving a half-hour earlier (he thinks) and by taking
lunch with them they ought to be able to make a good three hundred
and fifty miles by dark; then they can use the truck cutoff where there
is less traffic on Sundays.

So when the holiday begins they set out, wife, husband, children, dog
and all, heedless of warnings and prophecies, heedless of previous experience with holiday traffic. Knowingly they plunge into the heavy stream
of cars, struggle with it hour after hour, dodge from highway to country
road and back again, sometimes going fast, sometimes slow, but rarely
stopping, and glancing only briefly at the scenery. When the holiday is
over they come home tired, out of temper and with little or nothing to
show for their journey. Nevertheless they are somehow glad that they
have gone, and they will go soon again. Common sense urged them to
stay home and take care of many long postponed household chores; the
bloody prognostications of the National Safety Council should have
frightened them. But instead they chose to yield to a deepseated urge to

[from *Landscape,* vol. 7, no. 2, Winter 1958–59]

escape from the city and everyday surroundings into the open country, and, as I say, they were not sorry they did so.

Why the Sunday motorist likes driving even under these conditions is a puzzle worth exploring; but as to the instinct which takes him away from home, it is too universal, too elementary, I should think, to call for much analysis. We all have it whether we indulge it or not. Far from being a product of the motor age, it is probably as old as urban existence itself, and many of us can remember a time when this weekend and holiday exodus had a different, less deadly but no less popular form.

I find myself recalling the days when the streetcar was the chief means of Sunday transportation out of the city. In America, to be sure, we have all but forgotten what the streetcar meant; its heyday is now a good forty years behind us. In Europe it still plays a very useful role, even though it is beginning to surrender its monopoly to the private automobile, the scooter and motorcycle and chartered bus. The finest flowering of the streetcar-borne Sunday exodus in Europe occurred, I think, sometime between the two World Wars. It was then that city people began to have a little more weekend leisure (and a great deal of unwanted leisure during the depression years) for excursions, but did not yet have money enough for cars or even motorcycles. In those days almost every city family of modest means spent its Sunday in that countryside which lay within walking distance of the last stop of a streetcar line. That, in fact, was the late 19th, early 20th Century equivalent of our contemporary two-hundred-mile drive in the country. It was a very important institution in its time.

The Pedestrian Sunday

Is it too early to look back with affection on those streetcar excursions and the country (or suburban) holidays which went with them? They were certainly not exciting by modern standards; they were repetitious and quiet, but I recall them as having the comfortable sameness of a long established tradition, without surprises, perhaps, but without disappointments. Early on Sunday morning in the silent residential streets (entirely empty of parked cars) you saw small family groups bearing hampers, knapsacks, footballs, tennis rackets, water-wings, walking sticks, folding stools, sometimes a rake or a shovel, standing at every streetcar stop. A half hour later, out where the paved street came to an end among isolated tenements and factory yards and carbarns and cemeteries, you saw them emerge from the streetcar and set off into the open. They had a choice of roads: some went to their own well-fenced-in vegetable plot where they had a minute house with a trellis, others took off for a favorite patch of forest or a river bank or a hillside, or a rustic beergarden with a view. They all vanished in no time. Throughout the day you met them strolling at a child's pace on grass-margined lanes or playing games in a clearing, or picnicking or making love or snoozing with the newspaper over their heads, or gathering wild flowers in the woods. They clustered like flies around public swimming pools or beaches, and toward evening they tended to gather in village cafes and restaurant gardens where they ate the food they brought with them and listened to a small band play popular music. Late into the night the streetcars were once again crowded on their journey into the center of town with drowsy men and women and sleeping children. Their knapsacks and hampers were as full as when they set out, for now they contained flowers, berries, mushrooms, herbs, bundles of twigs for kindling from the woods, and sometimes lard and eggs and butter from farmers. As usual the outing had been rewarding.

Not merely in the material sense, of course, though I have often thought that what the Sunday holiday makers brought back with them illustrated very neatly what the city ought to derive from its green surroundings. The real benefits were of a different kind; the holiday makers were relaxed after healthy exercise and healthy rest; they had enjoyed an easy sociability with strangers, they had heard music, they had revived their awareness of natural beauty, and their ancestral ties with the land. Above all they had known an emotion, vaguely religious in character, best expressed in the old-fashioned phrase of "being close to Mother Nature." I do not mean to imply that this experience of the outdoors was necessarily varied enough or intense enough or even long enough; I merely mean that what these families got from their Sundays included almost everything (in a small degree) that we want, or used to want, from rural nature—everything from food to esthetic pleasure to spiritual sustenance. When the time comes for us to draw up the inventory of all the contributions the old suburban Sunday made to our culture we will be astonished by its richness. It has directly inspired schools of painting and writing and music—from the popular Viennese song which evolved in Grinzing to Seurat's "Summer Sunday at La Grande Jatte." Untold minor scientific and artistic accomplishments came from the same abundant source; local botanical and geographical and historical descriptions, small books of nature verse, amateur sketches and compositions, now relegated to store rooms and second-hand shops, testify to what this custom meant. And the most wholesome benefit of all, I think, was the love of nature it instilled among city dwellers of every class from childhood on.

All this, it must be remembered, was drawn from a very small area around the city; the range of the excursionist in pre-motorized days rarely exceeded twenty miles. And yet, once we have duly appreciated the splendid results and the poverty of means, once we have compared the old Sunday excursion with its frenzied, unfulfilled contemporary equivalent, we have to ask ourselves in all honesty whether it is possible or even desirable for us to revert to that old order. Much more has happened to us than the advent of the automobile; we have learned to see the world differently even on our holidays; we confront the familiar setting in a new manner. Broadly speaking, the former experience of nature was contemplative and static. It came while we strolled (at three miles an hour or less) through country paths with frequent halts for picking flowers, observing wildlife and admiring the view. Repose and reflection in the midst of undisturbed natural beauty, and a glimpse of something remote were what we chiefly prized. I do not wish to decry the worth of these pleasures; none were ever more fruitful in their time; but the layman's former relationship to nature—at least as part of his recreation—was largely determined by a kind of classic perspective and by awe. A genuine sense of worship precluded any desecration, but it also precluded any desire for participation, any intuition that man also belonged. The experience was genuine enough, but it was filtered and humanized; it was rarely immediate.

Contemplation without Participation

We need to bear these qualities in mind if we are to understand why the Sunday excursion (and the experience of nature that went with it) began to pall about thirty years ago. For even while the tradition seemed to be flourishing with unabated vigor—in the decades between the wars, that is to say—a new attitude toward the environment, a new way of feeling, began to emerge on both sides of the Atlantic. The prevalence of the automobile in America helped confuse the process, but in Europe it was easier to follow. In both parts of the world it resulted in the rejection of conventional pleasures.

What happened was that people—mostly young people—began to tire of the Sunday streetcar excursion relationship to nature and to go off on their own. In the process they discovered or adapted a variety of new ways of entertaining themselves and of exploring the world. Skiing, long established as a means of winter travel in Scandinavia and among a few eccentrics in America and England, first became really popular when city holiday makers took possession of it and transformed it. Rivers, once admired for their romantic turbulence, were suddenly alive with *faltboots*. Mountain climbing had formerly been a highly professional (and highly expensive) sport involving several trained guides for every Russian grand-duke or English milord; it now became a favorite lone-wolf pastime for amateurs. Small sailing boats, even during the Depression, multiplied on the lakes of Northern and Central Europe. Less popular but no less esteemed was the sport of gliding. Bicycling and hiking, neither of them novelties, of course, began to involve greater and greater distances with youth hostels and camping sites to accommodate the travelers. Lastly in the eyes of many young city dwellers the motorcycle came to be a symbol not only of cheap transportation to work, but of freedom and adventure on holidays.

Since the last war the number of new sports has increased enormously, with America taking the lead in devising them: skin-diving, parachute-jumping, surf-riding, outboard motorboating, hot-rod racing, spelunking—a variety of outlandish combinations like water skiing and hot-rod racing on ice. Some are so new that there is no telling yet how significant they are; others are too expensive or complicated to be widely popular. In the sense of being "character building" or "body building," few of them can qualify as conventional sports; some of them even lack the competitive or spectacular element altogether. Nevertheless according to the etymological definition of a sport as "a turning away from serious occupations," they certainly belong. To the question of why they should have risen when they did sociologists are ready with answers. They represent (so it appears) rebellion against parental authority or a compensation for the monotony and security of modern industrial society. Reuel Denney, for instance, describes hot-rodding as an attempt to escape from the conformity in automobile design imposed by Detroit and its "status car." These explanations are good enough as far as they go, but I still do not see how we can interpret any human activity without some reference to its chosen setting; I do not see how we can discuss purely in sociological terms any sport which is obviously designed as a form of psychological exploitation of the environment. To put it more simply, when people choose to practice a certain activity out of doors we ought to assume that the outdoors is somehow important to that activity. As I see it, those who adopted those sports did so because they had had enough of contemplation, and of the old sublimities which a century of poets and painters and musicians had interpreted over and over again. They may have resented the persistent loyalty of their parents to these things, but subconsciously what they wanted was a contact with nature less familiar and less pedestrian in both senses of those words—a chance to experience nature freshly and directly.

Yes, but how to achieve this freedom? One way was by acquiring mobility, mobility not only for going to new places but for seeing them in a new, non-pedestrian manner. Mobility does not necessarily mean speed. By the end of the 19th Century speed was hardly a novelty to the Western world, and we were aleady boasting of the Age of Speed that lay ahead. But the sensation of speed that a previous generation enjoyed when it traveled by fast train must have been strangely akin to its enjoyment of nature: it was passive, detached, and I daresay respectful, for when you sat inert in an upholstered railroad carriage and were swiftly borne along a pair of rails to an entirely predictable destination you could not flatter yourself that you were taking a very active part in the proceedings. It was quite a different sensation, however, infinitely more exhilarating, when you could actually manipulate the controls yourself, choose your own course and destination and rate of progress. That is why I would put the dawn of the new era at a time when we began to devise individual means of locomotion. The airplane and automobile; of course! But until a very few years ago, outside of America and perhaps England and France, how many men had ever driven a car or flown a plane? The precursors of those inventions, if not in time at least in popularity, were those simple, easily controlled devices: skis, sailboats, *faltboots,* gliders, bicycles and motorcycles.

The Discovery of Mobility

In any case, the holiday makers who adopted these contrivances soon found that their weekend contacts with nature had become new and exciting. Why? Well, for one thing, skis, *faltboots* and the rest were ideally suited for traveling in uncharted and hitherto inaccessible landscapes. The *faltboot* avoided navigable streams in favor of "white" water; the skier found himself moving down mountain slopes where there was no trace of man to be seen. The glider explored a new element, and so in a sense did the sailor. The motorcyclist went farthest afield of all and sought out rough terrain and paths impassable for four-wheeled vehicles. Each of these sportsmen saw aspects of the countryside that he had never seen before. To a generation which had never strayed very far from home, particularly to the urban European, this topographical freedom was a revelation; to be able to desert the well-marked, well-traveled path, to leave rails and highways behind and to move swiftly at one's own free will across remote hills and valleys and rivers and lakes was a fundamental departure from the old Sunday walk; and when to this was added a series of physical sensations without counterpart in the traditional contact with nature, then I think we are justified in calling this experience of the environment revolutionary.

For in these new, more or less solitary sports there is usually a latent, not entirely unpleasant, sense of danger or at least of uncertainty, producing a heightened alertness to surrounding conditions. Without much experience, without the presence of others to help and advise, without a stock of traditional skill, the sportsman, whether on skis or aloft or in a boat or on wheels, has to develop (or revive) an intuitive feeling for his immediate natural environment. Air currents, shifts of wind and temperature, the texture of snow, the firmness of the track—these and many other previously unimportant aspects of the outdoors become once more part of his consciousness, and that is why mountaineering, even though it entails a very deliberate kind of progress, has to be included among these new sports. None of them, for one reason or another, allows much leisure for observing the more familiar features of the surroundings; the skier or *faltbooter* or motorcyclist moves too fast (the mountaineer with too great a concentration on technique) to study the plants and creatures which his father loved to contemplate by the hour. A considerable loss, no doubt; nevertheless the new style sportsman is reestablishing a responsiveness—almost an intimacy—with a more spacious, a less tangible aspect of nature.

An abstract nature, as it were; a nature shorn of its gentler, more human traits, of all memory and sentiment. The new landscape, seen at a rapid, sometimes even a terrifying pace, is composed of rushing air, shifting lights, clouds, waves, a constantly moving, changing horizon, a constantly changing surface beneath the ski, the wheel, the rudder, the wing. The view is no longer static, it is a revolving, uninterrupted panorama of 360 degrees. In short the traditional perspective, the traditional way of seeing and experiencing the world is abandoned; in its stead we become active participants, the shifting focus of a moving abstract world; our nerves and muscles are all of them brought into play. To the perceptive individual there can be an almost mystical quality to the experience; his identity seems for the moment to be transmuted.

The discoveries of science and in particular the insights of artists and architects have made us familiar with changing concepts of space and matter and motion; without always understanding the theories we accept them as best we can. But what is our reaction when the man in the street tries in his own way to explore the same realm? We profess sympathy with the uncertainty, the inability to communicate of the contemporary

artist; why do we express little or none for the hot-rodder and his col-
leagues? Because his unconventionality comes too close to home; the artist and the physicist can be left to themselves (or so we think) whereas
the more modest variety space-explorer lives next door, and what we notice in particular about his activities is the rubbish-strewn landscape, the
disregard of time-honored esthetic values, the reckless driving. Still, even
these things should not blind us to the fact that the world around us, for
the first time in many generations, is being rediscovered by the young,
and being enjoyed. What will eventually come out of this headlong flight
into space we cannot as yet predict; for my part I see no reason why it
should not in time mark the beginning of a very rich and stimulating development in our culture.

Farewell to Mother Nature

Certainly here in America there can be no denying that the new attitude
is evolving with bewildering speed, and producing fascinating forms
worth studying for their own sake. The search for some contact with abstract nature is if anything more strenuous here than anywhere else. The
European sportsman-excursionist still derives a great deal of inspiration
from traditional landscape features: the picturesque village, the prosperous countryside, the glimpse of an older way of life. His American counterpart, on the other hand, seems increasingly bored by such pleasures.
What he wants is the sensation without preliminaries or any diversion;
what he wants is, in a word, abstract travel. Even the road is replaced by
an abstraction, starting and ending nowhere in particular in space—a drag-
strip in the desert or on a beach, a marked course on a snowy mountain
slope, a watery path between flags and buoys. And just as the contemporary artist or architect tends to simplify his medium to its essentials, to reduce the masonry elements in order to increase the flow of space, the
hot-rodder strips his car to a nubbin, the diver divests himself of his
heavy suit, the boat becomes little more than a shell. For the more drastically we simplify the vehicle (or the medium) the more directly we ourselves participate in the experience of motion and space. One feature of
the familiar world after another is left behind, and the sportsman enters a

world of his own, new and at the same time intensely personal; a world of flowing movement, blurred light, rushing wind or water; he feels the surface beneath him, hears the sound of his progress, and has a tense rapport with his vehicle. With this comes a sensation of at last being part of the visible world, and its center.

How general is this sort of experience? The answer, I suspect, is that it is not general at all, that it is confined to a very small minority. But this does not mean that there is not a very widespread interest in it, nor an equally widespread desire to participate. "The universal activity of racing sports cars is growing rapidly," (I quote from a letter which recently appeared in the *Christian Science Monitor.*) "This growth is no doubt attributable to the fact that the winding, sometimes hilly, often beautiful road-courses offer a challenge to many of us automobile drivers who must do 99 percent of our driving on relatively uninteresting highways and at restricted speeds. . . . The sports pages go on valiantly devoting half their space to trying to bolster attendance at the old sports instead of playing up contemporary interest in competitive automobiling (hot-rod, drag and sports car), skin-diving, water skiing, swimming, outboard-motorboating and racing, yachting, motorboat cruising, and flying . . . One reason for favoring these newer sports," (the writer adds) "lies in the element of participation . . . It may require some new reporting talent and some editorial policy changes to modernize our sports pages, but I believe the time for this advancement has long since arrived."

In other words, the change in sports has caught the sports editors napping. It could very well be that it has also caught some of our sociologists and recreationists napping just as soundly.

But to return to Average Citizen and his Sunday or holiday excursion. How, it will be asked, can he even remotely share in this new experience? Indeed, how can he know that such an experience exists? His free time is increasingly circumscribed by mechanized civilization and massed humanity. Each year his new car isolates him a little more completely from his surroundings no matter where he goes. As Paul Shepard wrote in an earlier number of *Landscape,* "The day is here when the air-conditioned automobile carries us across Death Valley without discomfort, without disturbance to our heat perceptors, and without any experience worth

mentioning." Nor does the design of our foolproof, sleep-inducing highways, or of our cars allow us to sense the surface under the wheels or to
feel the exhilaration of a steep climb, a sharp curve, or a sudden view.
We are compelled to move at a uniform speed, and we no longer even
have that earlier, Model T sense of participating in the functioning of the
automobile—one obvious reason for the popularity of European sports
cars. Like our grandparents, we are passively conveyed through a complex, well-ordered, admirable world—only now technology substitutes
for Mother Nature in distributing the bounty.

Even so, from time to time, Average Citizen catches a glimpse of a different kind of environment; brief, but enough to make him want to see
more of it. From the idiotically small window of a plane he manages to
see the wondrous, free, non-human, abstract landscape of clouds and lim-
itless sky; on a clear stretch of road, provided no state trooper is lurking,
he can step on the throttle and know the thrill of speed produced by his
own will; sun baths in the back yard give him a direct bodily contact
with air and light and sun, and on his vacation he sees the desert and the
open sea. Furthermore, his consciousness is constantly assaulted with the
new ideas of space and movement whenever he opens the morning paper
or looks at a specimen of modern art. His children wear space helmets
and addle their brains with science fiction and interplanetary comics.
The new world impinges on the old in even the best regulated of American homes. These are all fragments of a much wider experience, to be
sure, but in the long run they make him discontented with the familiar,
and drive him out onto the crowded Sunday highway in search of some
kind of release.

Participation through Movement

And that is where Average Citizen is still to be found: out on the crowded highway. How is he to be freed, I wonder, to discover for himself the
new reactions to nature, the new nature that awaits him? More highways?
Faster highways? Newer, simpler means of locomotion, newer and more
spacious sports areas, more remote vacation sites? Certainly no more
pretty parks or carefully preserved rural landscapes or classical perspec-

tives; limited though his choice is, he still has to be on the move one way or another, and he has to be made to feel that he is part of the world, not merely a spectator.

I confess, however, that this is a problem I am content to leave to others. The man who interests me is the excursionist or sportsman or part-time adventurer who has already found his way to the other world, and is already at home in that abstract preternatural landscape of wind and sun and motion. Because it is he, I think, who will eventually enrich our understanding of ourselves with a new poetry and a new nature mysticism. I would not go so far as to say that the Wordsworth of the second half of the 20th Century must be a graduate of the drag-strip, or that a motorcycle is a necessary adjunct to any modern "Excursion"; but I earnestly believe that whoever he is and whenever he appears he will have to express some of the uncommunicated but intensely felt joys of that part of American culture if he is to interpret completely our relationship to the world around us.

Library of Congress Cataloging in Publication Data
Main entry under title:
Changing rural landscapes.
A collection of articles from Landscape, 1951–1969.
CONTENTS: Processes and values: Jackson, J. B. Back
to the land. Sauer, C. O. Homestead and community on
the middle border. Wagner, P. America emerging. Jackson,
J. B. An engineered environment. [etc.]
1. United States—Rural conditions—Addresses, essays,
lectures. 2. Land use, Rural—United States—Addresses,
essays, lectures. I. Zube, Ervin H. II. Zube,
Margaret J., 1931– III. Landscape.
HN57.C485 301.35'0973 76–46599
ISBN 0-87023-228-2